Rights and
Responsibilities
of Participants in Networked Communities

Dorothy E. Denning and Herbert S. Lin, *Editors*

Steering Committee on Rights and Responsibilities
of Participants in Networked Communities

Computer Science and Telecommunications Board

Commission on Physical Sciences, Mathematics, and Applications

National Research Council

NATIONAL ACADEMY PRESS
Washington, D.C. 1994

NOTICE: The project that is the subject of this report was approved by the Governing Board of the National Research Council, whose members are drawn from the councils of the National Academy of Sciences, the National Academy of Engineering, and the Institute of Medicine. The members of the committee responsible for the report were chosen for their special competences and with regard for appropriate balance.

This report has been reviewed by a group other than the authors according to procedures approved by a Report Review Committee consisting of members of the National Academy of Sciences, the National Academy of Engineering, and the Institute of Medicine.

Support for this project was provided by core funds of the Computer Science and Telecommunications Board. Core support for the CSTB is provided by its public and private sponsors: the Air Force Office of Scientific Research, Apple Computer, the Advanced Research Projects Agency, Digital Equipment Corporation, the Department of Energy, IBM Corporation, Intel Corporation, the National Aeronautics and Space Administration, the National Science Foundation, and the Office of Naval Research.

STEERING COMMITTEE ON RIGHTS AND RESPONSIBILITIES OF PARTICIPANTS IN NETWORKED COMMUNITIES

iv

The National Academy of Sciences is a private, nonprofit, self-perpetuating society of distinguished scholars engaged in scientific and engineering research, dedicated to the furtherance of science and technology and to their use for the general welfare. Upon the authority of the charter granted to it by the Congress in 1863, the Academy has a mandate that requires it to advise the federal government on scientific and technical matters. Dr. Bruce Alberts is president of the National Academy of Sciences.

The National Academy of Engineering was established in 1964, under the charter of the National Academy of Sciences, as a parallel organization of outstanding engineers. It is autonomous in its administration and in the selection of its members, sharing with the National Academy of Sciences the responsibility for advising the federal government. The National Academy of Engineering also sponsors engineering programs aimed at meeting national needs, encourages education and research, and recognizes the superior achievements of engineers. Dr. Robert M. White is president of the National Academy of Engineering.

The Institute of Medicine was established in 1970 by the National Academy of Sciences to secure the services of eminent members of appropriate professions in the examination of policy matters pertaining to the health of the public. The Institute acts under the responsibility given to the National Academy of Sciences by its congressional charter to be an adviser to the federal government and, upon its own initiative, to identify issues of medical care, research, and education. Dr. Kenneth I. Shine is president of the Institute of Medicine.

The National Research Council was organized by the National Academy of Sciences in 1916 to associate the broad community of science and technology with the Academy's purposes of furthering knowledge and advising the federal government. Functioning in accordance with general policies determined by the Academy, the Council has become the principal operating agency of both the National Academy of Sciences and the National Academy of Engineering in providing services to the government, the public, and the scientific and engineering communities. The Council is administered jointly by both Academies and the Institute of Medicine. Dr. Bruce Alberts and Dr. Robert M. White are chairman and vice chairman, respectively, of the National Research Council.

Preface

In 1990, the Computer Science and Telecommunications Board (CSTB) decided to conduct a strategic forum on the rights and responsibilities of participants in networked communities. The board was motivated by the observation that participation in electronically networked communities was, even then, growing by leaps and bounds, in environments including the Internet, commercial network service providers, local bulletin boards, and company- and/or office-based networks.

In November 1992, a small invitation-only workshop was held in Washington, D.C., for prominent researchers and policy analysts to explore some of the issues that have arisen in this area; much of the background information in this report is drawn from that workshop. Participants in the workshop examined user, provider, and other perspectives on different types of networked communities, including those on the Internet, on commercial information services such as PRODIGY and America OnLine, and on grass-roots networks (e.g., those based on home electronic bulletin boards). Addressed were such questions as:

- What policies, laws, regulations, or ethical standards apply to the use of these services; who sets them; how are they developed; and how are they enforced?

- What are users' expectations regarding privacy and protection of other proprietary interests?
- What are the rights, responsibilities, and liabilities of providers or operators of these services?
- What are the rights, responsibilities, and liabilities of users of these services?
- What problems arise from connecting systems offering these services to systems that operate under different policies?

The forum, held in February 1993, had a somewhat different structure and aim. Although many of these same issues were addressed, the forum was organized around a set of hypothetical scenarios designed to illuminate how issues related to and associated with free speech, electronic vandalism, the protection of intellectual property interests, and privacy might emerge. The intent was to focus primarily on the concerns that policymakers in government and the private sector might have. As a result, much of the forum discussion involved questions of law and how the current legal regime helps to define the boundaries of what is or is not acceptable conduct on electronic networks.

The themes of the forum were heralded in a keynote speech by Congressman Edward Markey, chairman of the House Subcommittee on Telecommunications and Finance of the House Committee on Energy and Commerce. He noted the technological convergence of computer, communications, and entertainment technologies and pointed out that historical approaches based on differentiating these technologies may create problems for policymakers in the future. He underscored the importance of fundamental human values even in the new electronic medium of networks, and he argued strongly that policymakers have to address the negative as well as the positive aspects of the new medium.

This report is based on material drawn from the November 1992 workshop, the February 1993 forum, deliberations of the steering committee, and other material and events that have appeared in the interim. The workshop of the American Association for the Advancement of Science and the American Bar Association, "Legal, Ethical, and Technological Aspects of Computer and Network Use and Abuse," held on December 17-19, 1993, was particularly germane. The 1993 forum provided some background material on technology, legal underpinnings, and the then-current policy environment, much of which is incorporated into Chapters 1 through 3. Chapters 4 through 7 are devoted primarily to discussions of the scenarios. Chapter 8 focuses on the deliberations of the steering committee, although comments from other speakers and participants are liberally included. The five appendixes provide information about network technology (Appendix A), the agendas for the workshop and forum (Appendixes B and C), Mr. Markey's keynote speech (Appendix D), and biographies of the steering committee (Appendix E).

As the report of a workshop/forum event, this report does not attempt to draw conclusions, find definitive answers, or make specific recommendations; rather, its purpose is to illuminate, to question, and to articulate thorny and problematic issues that arise in this domain, thus helping to lay a foundation for more informed public debate and discussion. Where possible, CSTB has checked with individuals quoted in this report to ensure that their quotes are used in context. The steering committee and editors are responsible for the synthesis and analysis contained in this report. In addition, Chapter 1 and Appendix A are based in part on remarks made by Mitchell Kapor at the February forum, while Chapter 3 draws heavily on Anne Branscomb's presentation to the forum and her subsequent work. Finally, Laura Ost, an independent writer, developed the initial drafts of Chapters 4 through 7, and James Mallory of the CSTB staff contributed to Appendix A. The comments and criticisms of reviewers of early drafts of this report and of its anonymous reviewers are gratefully acknowledged.

The CSTB will be glad to receive comments on this report as well as any suggestions for further work in this area. Please send them via e-mail to CSTB@NAS.EDU, or via regular mail to CSTB, National Research Council, 2101 Constitution Avenue NW, Washington, DC 20418.

Contents

xi

Executive Summary

Electronic networks are a new communication medium that allows people to interact, coordinate action, and access and exchange information, all from their desktop computers. The networks have spawned a growing set of services that now include electronic mail, electronic publications and bulletin boards, conferencing, on-line information services and digital libraries, electronic transactions, and computer playgrounds.

Computer and communications technology can, with a high degree of assurance, be assumed to be increasingly capable for a long time to come. But such technology is an enabler for a variety of social phenomena that are more difficult to predict or understand, and the true intellectual challenges are much more likely to arise from people's use of networks for communication and information exchange than from the development of the technology to move large amounts of electronic information rapidly from one place to another. At a workshop held in November 1992 and a public forum in February 1993, technologists, service providers, policy analysts, lawyers, and social scientists from academia, industry, and government met to discuss some of the social issues raised by the emergence of electronic communities. This report is based on the discussions of the workshop and forum, as well as deliberations of the steering committee and material that has appeared in the interim. Its purpose is not to draw conclusions, find definitive answers, or make specific recommendations; rather, its purpose is to illuminate, to question, and to articulate thorny and problematic issues that arise in this domain, thus helping to lay a foundation for more informed public debate and discussion.

The communication and information-interchange aspects of electronic networks that provide benefits to our communities give rise to questions related to the rights and responsibilities of participants in those communities: Who is liable when someone posts a defamatory message, child pornography, or copyrighted material on a public bulletin board? What are the legal and ethical obligations of a service provider to screen public postings? What is the provider's obligation to protect the privacy of users of its services? Does responsibility flow from whether the provider has the technical ability to exercise control or from whether the provider chooses to exercise control? What is the role of regulation and the law versus that of ethics, informal community behavior, and the marketplace? What constitutes fair use of copyrighted information? What is the nature of informed consent relative to providing information?

The workshop provided a variety of background perspectives on issues such as free speech and privacy. The forum itself began with presentations on the nature of electronic networks and on the relevant legal perspectives. Within the current legal regime, creators of information are provided legal protection (and restraints) through copyright and patent laws. Publishers are protected primarily under the First Amendment, although they, too, must abide by relevant intellectual property law. Distributors govern their relationships with their sources through contract and with their customers through both contract and more informal business practices and codes of conduct. Carriers are subject to an elaborate regulatory system established by law and administered by the Federal Communications Commission and state regulatory agencies. Users are governed by social customs (commonly called "netiquette"), by contract with the providers of the services they use, by federal and state statutes, and by common law if they want to litigate about some harm that has occurred.

To explore the concepts of rights and responsibilities more fully, panels of experts at the February forum considered four areas: free speech, electronic vandalism, intellectual property interests, and privacy. For each area, two scenarios were presented and the panelists were asked to address relevant issues in the context of the scenarios; audience reaction to each of these panel discussions was also sought. After all four panels had finished, the steering committee attempted to identify and synthesize key themes. Certain important points emerged from discussions in these areas, as described below.

- *Free speech.* Providers of information services, such as commercial networks, have some leeway in defining the services they provide and thus their legal obligations, although their latitude may not be entirely unlimited. (For example, under certain circumstances, it is conceivable that even a private information service would have responsibilities for access that are more traditionally associated with public forums.) Providers that assert the right to control the content of public traffic may be subject to a more stringent liability (e.g., for defamation) for that traffic than those that do not assert such a right. At the same time, economic considerations may influence commercial networks to assert a higher degree of control. Information services supported by public funds, operated by government, or otherwise deemed public cannot discriminate among users on the basis of their electronic communications for First Amendment reasons; common carriers cannot refuse to carry traffic arbitrarily. Service providers of all types are well advised to establish the rules under which they provide their services, preferably in advance and perhaps in consultation with their users.

- *Electronic vandalism.* Current federal computer crime statutes are limited in that they focus on a class of acts whose elements can be difficult to define (e.g., what should be the definition of an "unauthorized" user in a world of highly interconnected computers?) and leave unaddressed a range of other acts that society may well wish to criminalize or limit (e.g., harmful acts not now criminalized that are deliberately committed by insiders who have legal access). Ancillary issues such as determining the level of damage may also be problematic in any given case. The relative newness of electronic networks as a medium for human communication and behavior means that we as a society collectively lack a great deal of shared experience about what is and is not (or should not be considered) a crime. Legal precedents and ethical standards applied to this new medium are ultimately the basis for criminalizing behavior.

- *Intellectual property.* Although intellectual property is traditionally the domain of copyright, patents, and trade secrets, most of the discussion of intellectual property matters in a networked environment is related to copyright, and discussions at the February 1993 forum reflected this weighting. Copyrighted electronic materials are covered by fair-use provisions in copyright law, although the nature of the electronic medium with respect to the reproduction and/or alteration of information amplifies concerns that have been addressed in other venues (e.g., in the domain of photocopiers). The use of intellectual property can also be controlled through licenses that have essentially unlimited freedom to specify contractually the conditions of use. Nevertheless, many people see electronic networks as a threat that will dilute the value of their intellectual work.

• *Privacy*. Privacy rights are never absolute but rather are rights that society balances with the need or desirability for disclosure under various circumstances. In many cases, individuals can make their own decisions and choices regarding their privacy (e.g., by using electronic mail (e-mail) systems that conceal the identity of the mail sender or by encrypting their communications); however, that privacy may be illusory. For example, a network that transmits e-mail may require its users to acknowledge that all traffic is public, but its regular users may become habituated to those warnings and may not internalize them. Thus, what counts as a "reasonable expectation of privacy" on electronic networks may not be clear in any given instance, even if policies are made explicit in advance.

Why are these areas contentious? Analysis reveals substantive disagreements rooted in value judgments. For example, forum participants disagreed on the extent to which continuity with past precedents is desirable. Lawyers and policymakers (mostly but not entirely) tended to argue that rights and responsibilities in a new domain are inherently rooted in existing rules. The laws, norms, policies, and practices governing any technology or behavior are formed from precedents, often in response to conflicts that arise from customer or market demands, criminal charges, and civil lawsuits. This is particularly obvious with common law, since the resolution of court cases depends heavily on legal precedents. But the same is true of other types of rules. In addition, rules tend to originate in informal sources and then, over time, become codified, first through common law litigation and later by legislation. For electronic networks, other media including print, telephones, radio, and television have provided a rich set of precedents to draw on. At the same time, precedents do not always provide easy answers, in part because they are never identical to the case at hand and in part because multiple precedents, with conflicting rules, may apply in a given case.

By contrast, many technologists who have extensive experience in using electronic networks assert that with a new medium and a new form of human expression should also come new rules of social intercourse specially adapted to that new medium. For example, many providers of network services have established policies and rules for the use of their services, and a body of case law and legislative statutes is emerging specifically for electronic services. Thus, for these people, questions about rights and responsibilities are not just philosophical or even open to being weighed according to the rules developed for other media, but rather must be addressed in rules specifically formulated for electronic networks.

Other topics on which workshop and forum participants expressed differing value judgments included the following:

- The extent to which the government should regulate behavior on electronic networks,
- The role of the marketplace in influencing behavior,
- The value of sharing information freely versus keeping information proprietary or private,
- The need for law that specifically relates to behavior on electronic networks,
- The nature of informed consent relevant to providing information, and
- The disposition of individuals who express no preference or inclination regarding their putative rights on electronic networks.

These disagreements, and others not mentioned above, are heightened as the concerns they raise (and have raised in the past) are magnified through the use of networking technology. In the past, any apparent resolution of such concerns has come about not because the concerns have disappeared or various stakeholders have changed their minds, but because political compromises and the need to move forward have driven decision making. Thus, resolution very much depends on the circumstances of the moment, and solutions or approaches to these disagreements will inevitably be a blend of past traditions, present realities, and possible future directions. Networking technology reopens traditional debates largely because it threatens the relevant status quo; with new circumstances, new compromises become necessary, and thus the same fundamental questions need to be reexamined.

If this is true, the debate over social norms on electronic networks will not differ substantively from the debate over any other controversial social issue, although the sophistication of the technological understanding required may well make the debate a less informed one. This is not to say that the debates should not be taking place, but only that our hopes about what such debates can accomplish should be moderated. These debates will not resolve fundamental issues or even result in consensus, but they can serve an educational role, illuminating and illustrating issues and providing alternative visions of the future for the concerned public.

1

The Nature of Electronic Networks

A NEW MEDIUM FOR COMMUNICATION

Electronic networks, which are webs of electronically connected computers, are a new medium for communication (see Appendix A for a description of network technology). Consider:

- They offer new tools for interacting, coordinating action, and conducting transactions.
- They enable new ways of accessing, distributing, and sharing and exchanging information.
- They provide new ways of learning, working, and playing.
- They give rise to new communities of people with shared interests and concerns, and they generate new interest areas and new concerns.

Although electronic networks share many of the properties of other communications media (e.g., postal mail, telephones, radio, and television), they provide individuals and organizations with communication tools that are faster and more efficient than postal mail, less

---

NOTE: In this chapter, all quoted material that is not otherwise identified originated with the individual noted, speaking at the November 1992 workshop.

controlled than radio and television, and more capable than the telephone or stand-alone fax machine of supporting large distributions. At the same time, networking (especially with networks tied to other networks) quite often involves multiple entities that may have different operating policies and procedures; in this regard, networking is unlike using more traditional media that tend to operate under more uniform and centrally formulated policy guidelines.

Using electronic networks, people share experiences and activities that bind them together (Box 1.1). Although long-distance transportation systems, telephones, and broadcast media have led to the formation of many geographically dispersed communities that are defined more by shared interests than geographical location (e.g., professional societies and nationwide clubs), networks provide a medium that transcends distance and further enriches the possibilities

---

Box 1.1
Some of What People Can Do with Networks

- Write an electronic message and send it through an electronic mail system.
- Receive and reply to an electronic message, forward copies to other people who are connected to the network, and clip out parts of the message for inclusion in a report.
- Browse the catalogue of a digital library and scan the contents of abstracts or full documents, and transmit selected documents to one's computer for later reading or for incorporating into other documents.
- Locate, download, install, and run software in network-accessible software libraries.
- Compose an article or newsletter and send it out to thousands of people on an electronic distribution list.
- Join on-line discussion groups that bring together people with shared interests, composing and reading messages that form a continuing conversation among potentially tens of thousands of "fellow travelers."
- "Telecommute" to work from a personal computer and engage in business with fellow co-workers, customers, and suppliers.
- Enroll in a program of study at a remote school, and then receive assignments, submit work, and interact with faculty through the network.
- Scan consumer catalogues and order goods and services.
- Connect to on-line entertainment centers and join other people in an electronic fantasy game.

for global communities. The total number of network users in the world is impossible to determine, but it would not be unreasonable to expect that tens of millions of people have some direct contact with networked connections.

The concept of a community whose existence is enabled by electronic networks is a useful departure point for the main topic of discussion of this report: the rights and responsibilities of the members of networked communities. Such matters inhere in how various communities of people use network technologies, and not in the particular technologies themselves, a point that is all too often lost in such discussions. The purpose of this report is to identify and address important questions related to free speech, privacy, intellectual property, and electronic vandalism as they arise in the context of networked communities; these four areas are not the only dimensions of the rights and responsibilities of network users, but they are important ones. This chapter and the next attempt to identify the characteristics of the new communications medium and of the communities that have formed around it.

THE NETWORK SCENE

The phenomenal increase in the number of network users in the past few years has been driven by two technologies: personal computers (introduced in the early 1980s) and networking technology (e.g., local area network technology introduced in the mid-1980s and further development of wide area network technology, including that deployed on the Internet and on various private and public data networks). In particular, Internet use has increased exponentially since 1983. By early 1994, estimates of the number of Internet users ranged from 2 million to 20 million (the upper end refers to electronic mail (e-mail) users), and by mid-1994 at least 3 million host computers were linked by the Internet.[1] The Internet is of particular interest because of its openness, the diversity of its users, and its characteristic use as a vehicle for experimentation with new information and communications services.

Two fundamental aspects of the Internet are decentralized: its technology and its governance. The Internet is based on packet-switching technology that transmits a message between two points by breaking the message into packets that travel independently and

---

[1]Press release of the Internet Society, August 4, 1994. The Internet Society is the international organization for the Internet, its technologies, and applications.

often through different routes between sender and receiver. As with telephone traffic, the path taken by data sent through the Internet is often not known in advance.

The governance of the Internet is also decentralized. That is, each site on the Internet operates under its own locally formulated code of behavior or conduct, though each site communicates with other sites. This is especially relevant given that the Internet accommodates host sites of different types (academic, commercial, government) and host sites located in different regions of the country or in different countries. Since message traffic from point A to point B may traverse sites with codes of behavior or conduct that are very different from those that govern point A or point B, the retransmission of certain types of message traffic along intermediate sites may be regarded as a violation of the code of those intermediate sites. As David J. Farber, a University of Pennsylvania professor of computer and information science, put it, "I don't know what they are doing with my traffic. I don't know what laws I'm violating . . . . [W]e don't seem to have either international agreements or case law [that are known to the general community] that apply to the electronic communications area, and that is going to be more and more a serious problem." International implications may also differ depending on whether the network(s) in use is public or private.

Future developments may make today's Internet pale by comparison. Present expansion of the Internet community has been fueled by an increasing number of computer-literate individuals. A much larger pool of potential networkers is represented by the canonical person on the street, who will resist the use of computers until they are as easy to operate as telephones or televisions. This is a market that telephone companies and cable television companies hope to tap. Start-up companies, with backing from major consumer electronics companies, are trying to bring messaging and information services to the general public, as are a number of larger, more-established companies such as the Prodigy Services Company and CompuServe. The promise is sophisticated, easy-to-use, user-friendly information services, not for 10 million people but for hundreds of millions of people.[2]

An important political development is the Clinton administration's

---

[2]The issue of access to electronic resources by those in low-income, rural, and inner-city areas is an important question that many commentators and analysts are addressing under the question of what "universal access" means in a networked environment. For reasons of time and resources, the steering committee chose not to address this point in detail.

support of a comprehensive network complex referred to as the National Information Infrastructure (NII). The administration's view of the NII is sweeping and grand and promises significant benefits for all in U.S. society.[3] But if the NII is indeed to contribute to the betterment of society, it behooves the nation to address issues related to the behavior of users and providers, as well as questions of the NII's extent and reach, capability, quality, timing of service availability, and cost.

The broadening base of users of information services, the contemplation of new uses in inherently personal arenas (e.g., the delivery of social services), and the growth of information services for businesses and consumers have been bringing the issues of rights and responsibilities to the attention of policymakers at state and federal levels. Typically, the motivation for policy discussion is some kind of a problem: a security breach in a private or public network, the harassment of an individual via electronic mail, the controversy over whether material available over a given service is pornographic and whether access to such material can or should be controlled. Enough perceived problems and controversy have been generated that private guidelines (codes of conduct) are being developed, tried, and revised, and public guidelines (including regulations, advisories from the Office of Management and Budget, or laws) are being discussed.

---

[3]The following quotes are taken from the administration document *The National Information Infrastructure: Agenda for Action:*

> All Americans have a stake in the construction of an advanced National Information Infrastructure (NII), a seamless web of communications networks, computers, databases, and consumer electronics that will put vast amounts of information at users' fingertips. Development of the NII can help unleash an information revolution that will change forever the way people live, work, and interact with each other:
>
> • People could live almost anywhere they wanted, without foregoing opportunities for useful and fulfilling employment, by "telecommuting" to their offices through an electronic highway.
> • The best schools, teachers, and courses would be available to all students, without regard to geography, distance, resources, or disability.
> • Services that improve the U.S. health care system and respond to other important social needs could be available on-line, without waiting in line, when and where they are needed.

See Information Infrastructure Task Force, *The National Information Infrastructure: Agenda for Action,* Washington, D.C., 1993.

In the Clinton administration the focal point for policy consideration in this area has been the Information Infrastructure Task Force (IITF), a cross-cutting interagency group whose role is to explore areas where policy may need to be formulated or changed and to gather inputs from within and outside the government. A major component of that task force is the Information Policy Committee, which is examining policy issues in the area of privacy, security, and intellectual property protection. Activities of that committee through mid-1994 included outreach, input gathering, deliberation (resulting in a draft of a set of privacy principles that was circulated for public comment in May 1994), the planning of a public forum to obtain input on security issues (held in July 1994), the release in July 1994 of a report on how copyright law should be updated in the age of ubiquitous electronic networks, and the planning of a conference to address how fair-use provisions of copyright law may be applied to the electronic realm. Although some of the administration activities are intended to drive proposals for legislation, the Congress itself has addressed related issues (e.g., Senator Paul Simon introduced S. 1735, the Privacy Protection Act of 1994, a bill that would establish a U.S. data protection commission to advise the U.S. government on, among other things, matters related to protecting data stored in electronic form), and telecommunications reform legislation may eventually address relevant issues. Meanwhile, these issues are also being addressed—through reports, polemics, meetings, and exploratory committees—by an assortment of private entities, including trade, professional, and advocacy groups, such as the Telecommunications Policy Roundtable, the Coalition for Networked Information, EDUCOM, the Electronic Frontier Foundation, Computer Professionals for Social Responsibility, the Internet Society, the Computer Ethics Institute, the Information Industries Association, and so on, as well as direct representation from the entertainment, cable, telephone and telecommunications, and information-providing and publishing industries.[4]

This large and growing set of actors on the policy stage suggests that the future may be only dimly visible today. What is clear is that the nation is on the threshold of an era in which the networking

---

[4]Several groups have issued reports in the past year with the intention of influencing public policy and attitudes regarding the NII. See, for example, Computer Professionals for Social Responsibility (CPSR), *Serving the Community: A Public Interest Vision of the National Information Infrastructure,* CPSR, Palo Alto, Calif., 1994; and Council on Competitiveness, *Competition Policy: Unlocking the National Information Infrastructure,* Washington, D.C., December 1993.

environment will be highly open and commercial, very heterogeneous, and with rules that may look quite different from those to which "old hands" at networking are accustomed. The ultimate social impact of electronic networks may be about as well understood today as that of the telephone in 1876. Many of those who first supported the deployment of the telephone argued that it would enable music to be brought from the performance hall to the home; many fewer imagined that it would become the center of personal and business communications.

NETWORK METAPHORS

Electronic networks offer new communication tools for interacting and for coordinating action. Three metaphors have emerged for characterizing the new medium and the functions it enables: cyberspace, the information superhighway, and the electronic marketplace. Each of these metaphors emphasizes a particular use of electronic networks and, in so doing, facilitates our understanding of the new medium by illuminating its salient features and the social and legal issues they raise. As with most metaphors, however, the interpretations that emerge are limiting and can be misleading—a point often overlooked by their users. Thus, this report does not systematically favor the use of one metaphor over another.

Cyberspace

The term "cyberspace" originated with the science fiction writer William Gibson, who characterized it as a "consensual hallucination of visually realized data achieved through plugging in to a global computer network."[5] It could be experienced by making a direct, physical link between an individual's brain and the semiotic data available on a global computer network. The use of the word "space" within the term suggests that cyberspace has some similarities to physical space.

Current usage focuses on this notion of electronic networks constituting a virtual place that transcends physical space and national boundaries.[6] People go to cyberspace, travel on its electronic roads

---

[5]William Gibson, *Neuromancer*, Ace Science Fiction, Berkeley Publishing Group, New York, 1984. Note that cyberspace need not be limited to visually realized data; simulated sounds and touch are also included.

[6]Although the idea of nonphysical space may sound metaphysical, John Perry Barlow has observed that most individuals have already had some experience with cyber-

and highways, and settle into virtual communities—all from their office or home. They visit "electronic pubs" and "town squares" by participating in discussion groups on bulletin boards and "newsgroups." They join virtual communities that form around shared interests. They socialize with people all over the world.

The metaphor is used to suggest a new frontier—a place that is not yet settled, where anarchy often prevails. Issues related to rights and responsibilities are often framed in the context of how to civilize or establish law and order in the new frontier. Inhabitants believe they should be free to create their own rules and laws. The metaphor also draws attention to issues relating to freedom, particularly free speech and free press, and raises issues of privacy and of search and seizure within the virtual space.

A potential limitation of the cyberspace metaphor is that it may overemphasize a new virtual space at the expense of the broader and historical context in which networking technology is embedded. Other communications technologies (e.g., mail, telephones, television) enable people to transcend the limitations of space and time as well, and considerable human effort has been expended to establish rules and practices for using these technologies. The focus on a new frontier overlooks the societies that developed and deployed the new technologies and the laws, practices, and traditions these societies have already established governing behavior performed by someone within their jurisdiction. The metaphor suggests that we are engaging in some wholly new set of actions when in fact we are merely using new tools to engage in very basic human acts of communication and coordination of action.

The Information Superhighway

The "information superhighway" metaphor emphasizes the use of electronic networks to access and distribute information. It suggests that the primary purpose of the networks is to carry information and that that information travels over the information highway much as cars and trucks travel over physical highways. The information highway links information providers (e.g., digital libraries) to users. On/off ramps connect a high-bandwidth superhighway to

---

space as a practical reality: cyberspace is the location of the conversation when two or more people talk on the telephone. Any money that you do not currently have in your wallet or purse or bank vault, if it exists anywhere, exists in the cyberspace of the modern electronic banking system.

lesser highways, and lesser highways to organizations and individuals. The metaphor is embodied in the administration's plans for the NII.

The focus on information leads to the issue of who has access and whether the new technology will lead to a widening gap between information "haves" and "have nots." The metaphor also raises issues relating to intellectual property and to privacy and security (information protection). Publication and distribution of information raise issues of free speech and free press, and the possession of information raises issues related to search and seizure.

A particularly controversial implication of the metaphor is that it suggests a key role for government in promoting its development: just as the government was a major force behind the construction of the federal highway system, so also will the government be a major force driving the deployment of the information superhighway. The controversy has several important aspects. The first is that it suggests a major government role in funding deployment, in spite of the fact that the administration has made it clear that its vision for the NII depends on the private sector making most of the investments for the physical infrastructure and for the services and information to be provided over that infrastructure. In the administration's view, the role of government will be to provide leadership and guidance for the NII. A second important aspect is that many proponents of the information superhighway argue for a bottom-up, grass-roots character to its development and deployment, while the construction of the interstate highway system was undertaken top-down with intimate government direction. A third aspect is that major government involvement suggests the possibility that federal, state, and local governments will be working together to provide the network of connecting streets, on-ramps, off-ramps, and the like to every community on the main "superhighway." Whether such partnerships will in fact be feasible remains to be seen, and the nature of appropriate government involvement in the National Information Infrastructure remains the subject of considerable argument and debate.

Finally, a potential limitation of the metaphor is that it may overemphasize information at the expense of the use of the technology to coordinate action, to conduct transactions, and to communicate informally. For example, many of the examples offered by proponents of the information superhighway involve the transmittal of large volumes of information, information that is traditionally understood as something related to knowledge and data or something that can be learned from a book or a lecture. Many fewer examples deal with the use of the information infrastructure for coordination and the like, despite the models offered by many service and manufacturing enterprises in using private networks (perhaps operating over leased public lines) for these purposes.[7]

The Electronic Marketplace

For many people, the "electronic marketplace" metaphor emphasizes the growing use of electronic networks to engage in transactions, both formally and informally, to acquire goods (often information) and services such as education. It resembles the physical marketplace and is a part of the global marketplace.

Because information is an important good in the electronic marketplace, the marketplace metaphor raises the same issues as the information superhighway metaphor. However, it does so more powerfully by showing that the issues arise in the context of transactions. For example, the information superhighway metaphor suggests that it is sufficient to give everyone access to the electronic highway in order to close the gap between the information "haves" and "have nots." The marketplace metaphor shows that simple access may not make much difference if the information available through the technology must be purchased. The marketplace metaphor also shows that information privacy relates to the use of information in transactions (e.g., selling mailing lists) and not just to the information independent of its use. In addition, the marketplace metaphor raises new questions about concerns such as accountability and fraud, racketeering, and other criminal acts associated with transactions. Finally, it calls attention to the idea that while individuals may have the right to market information goods and services, the costs of selling their wares on networks may be prohibitive.

The marketplace metaphor is sometimes referred to as the "electronic agora" in analogy with the public square in ancient Greece that served as the marketplace and center of civic life. This usage brings to life the community aspects of the cyberspace metaphor, but does so more powerfully by including the transactions that take place within those communities.

The use of the marketplace metaphor is likely to increase as more

---

[7]See, for example, Computer Science and Telecommunications Board, National Research Council, *Information Technology in the Service Society: A Twenty-First Century Lever*, National Academy Press, Washington, D.C., 1994; Computer Science and Telecommunications Board, National Research Council, *Information Technology and Manufacturing: An Agenda for Research*, National Academy Press, Washington, D.C., forthcoming.

goods and services become available over networks and methods of making and receiving electronic payments securely are implemented. This metaphor emphasizes the actions that people take using the new communications medium, and not just the inert information that flows through the networks or the communities that form around common interests. Rights and responsibilities relate to actions and define which actions are encouraged, allowed, discouraged, or forbidden.

However, despite its strengths, the metaphor also has potential limitations. One is that it is easily interpreted to emphasize commercial transactions at the expense of noncommercial interactions. Thus, not-for-profit ventures and activities (e.g., education, government) may feel less than comfortable with it. A second is that with its economic connotations, it suggests that a kind of laissez-faire philosophy governs operations in the electronic marketplace, when the extent and nature of government regulation are policy issues that have not yet been settled (and are unlikely to be settled for a long time to come). A third is that some commentators use "electronic marketplace" with closer attention to its political origins and in the same sense that the term "marketplace of ideas" is used—as a description of an arena in which ideas and/or information are freely circulated to all without regard to size or political power, are exposed to public scrutiny or use, and survive or disappear according to their quality and value.

2

Networks and Society

The glamour of network technologies is impressive, but it should not blind us to the social and intellectual challenges posed by new media. The real challenges are much more likely to arise from the people using networks for communication and information exchange than from the development of the technology to move large amounts of electronic information very rapidly from one place to another, though technology itself may enable new ways of meeting these challenges. It is the communication and information interchange aspects of the networks that provide benefits to our communities and also give rise to the need to examine how networks relate to culture and society.

NETWORKS AND CULTURE

Networks and culture can be examined from two perspectives: the nature of network culture (culture "in the small") and the impact of networks on culture at large.[1]

NOTE: In this chapter, all quoted material that is not otherwise identified originated with the individual noted, speaking at the November 1992 workshop.

[1]The literature on the social dynamics in networked environments is extensive. Good discussions include: Lee Sproull and Sara Kiesler, *Connections: New Ways of Working in the Networked Organization*, MIT Press, Cambridge, Mass., 1991; National Research Council, *People and Technology in the Workplace*, National Academy Press, Washington, D.C., 1991; and Tora K. Bikson and J.D. Eveland, *Technology Transfer as a Framework for Understanding Social Impacts of Computerization*, The Rand Corporation, Santa Monica, Calif., 1990.

It is clear that members of networked communities engage in social behavior. For example, these individuals have been known to "meet" each other in these on-line communities. At these meetings, they discuss personal and professional matters, find jobs, advertise their services, engage in "stalking" behavior, and become visibly angry.

On-line social interactions provide the behavioral basis on which much network culture is generated and reproduced; such interactions are also the subject of the behavior that is regulated by culture at large. As a general rule, culture can be regarded as a collection of three elements: (1) values, or general statements about the "desirable"; (2) behavioral norms, including etiquette and convention, that provide the basis for judging, in individual situations, whether a value is being observed, thus regulating behavior in the interest of implementing or reinforcing values; and (3) beliefs that provide history, mythology, and world view. All of these apply to networked communities in various forms. For example:

1. Many networked communities believe strongly in the desirability of unfettered communication. At the same time, many (especially those that reside on the Internet) believe strongly that commercial advertising traffic is highly undesirable.

2. People observe forms of etiquette and social convention in networked communities. For example, behavior that is regarded by community members as antisocial is censured through peer pressure and comments (but generally not censored). Even behavior as subtle as typing in capital letters has significance, in that TYPING IN CAPITAL LETTERS IS WIDELY REGARDED AS THE ELECTRONIC EQUIVALENT OF SHOUTING.

3. Networked communities have their own myths, their own equivalents of urban folklore stories such as alligators in the sewers. One persistently recurring myth is the report that some government agency is about to impose a modem tax that will make it prohibitively expensive for private users to use telephone lines to transmit data. The story may have some origin in truth,[2] but it surfaces so

---

[2]This story apparently has its roots in two distinct events. In the early 1980s, a proposal was circulated to impose a charge on the leased lines used by on-line information services providers. The providers would have paid this charge, but in all likelihood it would have been passed to the ultimate consumers. In addition, in the early 1970s, a proposal surfaced in the Midwest to impose a small additional surcharge on telephone lines that were self-declared to be supporting data communications traffic. Both these proposals died very quickly, but their effects linger on.

often as a "new" story that it can be regarded as the network equivalent of an urban folktale.

In short, network cultures exist because networked communities exist, and they are as diverse as any collection of human beings can be. Networked communities exist because networks enable individuals with shared interests or affinities to affiliate electronically.

As with physical communities, network cultures are not homogeneous, and the participants in networked communities often have conflicting values. A recurrent issue in university communities, for example, is whether pornographic images or "hate speech" should be banned. Although some universities have banned entire electronic "newsgroups" that were judged pornographic, most have stayed away from such censoring.[3] Several years ago, Stanford University banned an electronic newsgroup on humor because of complaints about an ethnic joke. The newsgroup was subsequently restored after campus protests about free speech.

Electronic networks have also had an impact on culture at large. For example, commercial on-line services have become the virtual equivalent of "singles" bars for many individuals. Using the real-time conferencing services provided by these services, many people seek on-line conversation with potential romantic partners. They flirt through their keyboards, developing romances and carrying on "illicit" or clandestine affairs. In some instances, individuals who have interacted on-line meet face to face and develop relationships in person; marriages have been known to result from such interactions.

A second example is that computer networks have become a grass-roots vehicle for lobbying against government and commercial actions. For instance, a large software company once attempted to market a product that contained a large database of information about consumers. An electronic protest was organized against it on the grounds that it had many privacy implications. These protests flooded the electronic mailbox of the company's chief executive officer. As a result, the product was withdrawn.

Finally, computer networks have served as important political tools. For example, the Relcom, the Soviet network on the Internet, played a significant role in the dissemination of information during the Soviet coup attempt in 1991. Relcom was used both for communication among the coup resisters and for communication with the

---

[3]A newsgroup is one form of electronic conference. See Appendix A for more details.

outside. Some considered it to be one of the best sources of information.

By themselves, the examples given above and many others like them are not individually significant; that is, they do not necessarily demonstrate how networks will be used in the future. But taken together they do illustrate a general and fundamental point—that networks may change conduct and behavior in significant and multiple ways across life and society.

CONFLICTING VALUES IN NETWORKED COMMUNITIES

As noted above, networked communities are not monolithic. With their proliferation (and the interconnection of many across geographic or political boundaries), electronic networks have been described variously as models of "anarchy," "extreme decentralization," and "accelerating decentralization." Most networked communities have little hierarchical administrative structure and few central administrative procedures or systematic methods of policy enforcement, although some communities (e.g., those resident in universities or private companies) operate under the aegis of a single administrative entity and are subject to the rules promulgated by that entity—and yet each of these communities may manifest its own "culture" within those rules. Under these circumstances, it is perhaps understandable that different networked communities might have different values.

Many universities govern their electronic networks through campus policies that are substantially the same as the policies governing other pieces of university infrastructure. Freedom of speech (discussed more extensively in Chapter 4) tends to be an overriding value on these networks, based on the principles of academic freedom.[4] But incidents challenging this freedom arise frequently. Jeffrey I. Schiller, network manager at the Massachusetts Institute of Technology, for example, notes that at least once a month someone asserts harassment as the result of someone else's electronic free speech. The values that govern behavior may be blurred further by campus

---

[4]For instance, the "Joint Statement on Rights and Freedoms of Students" issued in 1967 by the American Association of University Professors, U.S. National Student Association, Association of American Colleges, National Association of Student Personnel Administrators, and National Association of Women Deans and Counselors holds that "academic institutions exist for the transmission of knowledge, the pursuit of truth, the development of students, and the general well-being of society. Free inquiry and free expression are indispensable to the attainment of these goals."

connections through the Internet to other institutions or organizations that may have different values. "And the only thing that has made the situation tenable has been the fact that most of the policies have significant amounts of overlap and nobody is enforcing them," Schiller said. "Only the most outrageous behavior will ever raise an eyebrow . . . . "

A different balance of values is found in the world of commercial network service providers. Commercial providers are highly motivated to provide a range of network services that appeal to large audiences and generally try to suppress message traffic found offensive by significant segments of these audiences. For example, a commercial provider will often explicitly prohibit overtly sexual real-time chats on its network.

A third example illustrates the problem of even defining the boundaries of an electronic community. In July 1994, a California couple was convicted by a Tennessee jury of transmitting obscene images through interstate telephone lines through their bulletin board system,[5] thus raising the question of whether it is the community standards of California or Tennessee that are relevant to the Supreme Court's position that community standards should define what does or does not constitute "obscenity."

In short, people who use electronic networks subscribe to a range of values that is highly diverse. Although they still represent an elite group in the context of the global population, users vary widely in terms of age, cultural background, and interests. For example, William Dutton, a professor at the Annenberg School for Communication at the University of Southern California (USC), polled all users of a USC bulletin board system and a random sample of 50 users from the Public Electronic Network (PEN) system[6] in Santa Monica, California, from among all those who had logged onto the system more than 10 times during a particular month. He found that users "vary dramatically in the values and norms they bring to how rights and responsibilities should be negotiated."[7] Based on this work, Dutton developed a typology of user values that includes five categories:

---

[5]"Couple Convicted of Pornography Sold Over Computer Network," *New York Times*, July 31, 1994, Section 1, p. 15.

[6]The PEN had more than 4,000 registered users as of November 1992, and it provides two-way information, e-mail, and conference services for citizens, free of charge.

[7]Dutton suggested several reasons for these differences, including characteristics of the network system used (particularly its ownership), individual differences among users, and characteristics of the social context within which the discussion takes place.

1.   Civil libertarians, who value free speech above all;

2.   Regulators, who "call for censorship and actually are more concerned about the privacy invasion and intrusive aspects of the system than they are about First Amendment issues";

3.   Formalists, who rely on previously written documents and are satisfied with the status quo;

4.   Property rights advocates, who believe individual users have no rights and are given the privilege of using the system by USC, which has the right to make the rules; and

5.   Balancers, who weigh the conflicting values of free speech and privacy.

To many experienced network users, this typology does not seem at all unfamiliar (though the various sizes of the groups differ depending on the electronic community in question). The fragmentation is heightened by the fact that user viewpoints are not immutable, especially as a given individual may be a member of many different communities operating on many different networks. Sara Kiesler, a professor of social sciences and social psychology at Carnegie Mellon University, noted that a user's attitudes may change because "this same person in a different situation has a very different view of the rights that he has."

ENFORCEMENT OF BEHAVIORAL NORMS IN NETWORKED COMMUNITIES

The means for enforcing behavioral norms in networked communities are as diverse as user values. For example, commercial providers have contractual agreements with users. Several million individuals use these services.[8] Users tend to view the contracts through which their use arrangements are governed less as disciplinary mechanisms than as "guarantees of quality of service," according to Eberhard Wunderlich, division manager of low-speed services product management at AT&T Data Communications Services. Nevertheless, these agreements prohibit certain types of conduct, such as the distribution of obscene or threatening material, and network providers are under pressure to enforce such rules to avoid offending clients.

---

[8]The three largest services are CompuServe (2.3 million users), PRODIGY (2 million), and America OnLine (1 million). Figures provided September 1994 by the public relations office of each organization.

Still other networked communities are governed by informal social pressures. For example, on the PEN, self-appointed "thread police" keep users from straying off the topic in conferences. "Anything goes, and, therefore, they push a great deal of responsibility on self-policing," explained Dutton. "And the thread police have been one of the creatures of the self-policing . . . . The thread police will either talk to people on-line or try to talk them into getting into another conference." On the Internet, miscreant behavior is often addressed by members of the community sending many electronic messages protesting the behavior to the person responsible.

Social pressures also may be applied on corporate networks. For years now, the IBM Corporation has had a "sensitive forum" where users may discuss problems concerning each other's messages; almost immediately after the forum was established, users began making rules, according to Davis A. Foulger, advisory programmer at IBM's T.J. Watson Research Center. "The rules were never written down formally as a code, but in fact they are enforced rigorously by e-mail, and nobody wants 10,000 IBMers to put mail in their mailbox all at once," Foulger said. "Dozens of these kinds of rules . . . have emerged through discussion in our media."

Yet another sort of etiquette is common on many electronic bulletin boards and computer conferences. Conferences normally are moderated by an individual who lacks formal powers but regulates activity by severing electronic links with troublesome users. As Jack Rickard, editor of *Boardwatch* (a magazine that serves bulletin board operators), said, "There is very limited control of the content of these message areas. . . . The bulletin board right now is essentially untaxed, unregulated, and censored by rumor more than anything else."

HOW CULTURAL NORMS EVOLVE:
AN HISTORICAL PERSPECTIVE

Cultural norms may derive from many different sources. As described by Professor Henry Perritt of Villanova University School of Law, these sources include legislatures, administrative agencies, negotiated contracts, tort law, common law, informal associations, common practice, and academia. Rules tend to originate in informal sources, and then, over time, are codified, first through common law litigation and later by legislatures, Perritt said. Network enforcers basically have three options: disconnect rule breakers, employ peer or social pressures, or apply the law. Although making rules for electronic networks is challenging, enforcing the rules may be even more problematic, he noted, because of technological and economic barriers to enforcement. Attorney Lance Rose gave the example that "the ease of spread of information on networks can, on a practical level, defeat various kinds of intellectual property enforcement. . . . If you put something on the network, it's gone before you blink an eye."

Alan F. Westin, a professor of law and public government at Columbia University, described a predictable pattern to the development of rules and norms for new forms of communication. At first, the activity is sheltered and informal, involving only a small number of individuals. As more users engage in the new form of communication, government agencies begin to seek related records for tax and criminal investigations. Then users begin to file civil lawsuits against various parties for assorted wrongs. Tax issues also arise; the government must determine, for instance, whether and how tax-exempt organizations may use the emerging form of communication. Then freedom-of-information issues come to the fore; some records will be public, especially if taxpayer money has been used, but someone must identify the exceptions that should be kept private.

Thus, the law becomes important because it settles conflicts and, in so doing, forms a context for resolving issues in the future. As attorney Allan R. Adler said, "What the law is supposed to do is prevent and resolve conflicts, typically over values and property . . . [and] any communications medium, by definition, is going to raise the prospect of conflicts over values and property." Electronic forms of communication are evolving in an established legal context and at the same time are reshaping the law, Adler noted.

Because the law is the forum of final appeal, the development of legal precedents or models that can be used in networked communities is of significant interest. This is no small undertaking. Information services once fell into three more or less distinct categories— newspapers, broadcasts, and telephone networks—each with different rules of access, liability, and freedom of speech. "Well, the problem is that these neat compartments are breaking down . . .," Perritt noted. "Not only that—we have some new kinds of roles emerging. It's no longer only the provider of the information and the customer; we [also] have new kinds of intermediaries that are very important."

Anne Wells Branscomb, a communications lawyer with the Center for Information Policy Research at Harvard University, said electronic networks "are a new environment with which we have not had enough legal experience to know exactly what rules apply. So I think we have a whole new legal world to worry about."

The emergence of cultural norms is often crystallized by some kind of dramatic event or crisis that underscores the need for new norms or the ambiguity in existing norms. For example, certain crime legislation in California passed in 1994 mandating life sentences without parole for those with three convictions for certain types of felonies was the fairly direct outcome of a social campaign initiated after a dramatic case of kidnapping and murder. More generally, the development of civil liberties law has been chronicled in large part by decisions precipitated by specific acts of injustice perpetrated by kings and political leaders in the context of uncertainty or ambiguity about their privileges or rights to commit those acts.

As of this writing, a similar pattern can be seen in the response of networked communities to the "Clipper" proposal. This proposal from the Clinton administration calls for a standard of secure communication that would nevertheless enable law enforcement authorities to decipher these communications upon issuance of an appropriate judicial warrant.[9] Although the Clipper proposal is limited to telephone communications, the overwhelming sentiment in various communities and public forums on the Internet is negative.[10] Congressional hearings have been called on this subject, and the status of legislation affecting this proposal is unclear.

ETHICS, LAW, AND THE PROMOTION OF SOCIALLY ACCEPTABLE BEHAVIOR

What is the role of law in the promotion of socially acceptable behavior? Law is arguably a norm that applies across the entire jurisdiction of a political state. Moreover, the phrase "socially acceptable behavior" in this context implies a consensus on what constitutes behavior that is acceptable across that jurisdiction. For a variety of reasons, some deny that a consensus exists, or even that a consensus can exist, and they do not recognize the value in consulting existing legal regimes for guidance. For example, David Hughes, of Old Colorado City Communications, claimed at the February 1993 forum that human behavior on electronic networks exists outside of

---

[9]Office of the White House Press Secretary, "The Clipper Chip," August 16, 1993; Office of the White House Press Secretary, document describing implementation of recommendations of the interagency review of encryption policy, February 4, 1994. Both documents can be found in David Banisar, *1994 Cryptography and Privacy Sourcebook*, Electronic Privacy Information Center, Washington D.C. (Diane Publishing, Upland, Pa.), 1994.

[10]Computer Professionals for Social Responsibility, "Computer Users Call on Administration to Drop Encoding Plan," press release of April 29, 1994. This document can be found in David Banisar, *1994 Cryptography and Privacy Sourcebook*, 1994.

present social norms and laws. He drew an analogy to space exploration, noting that "there was no space law until you could reach there. Space was not legally regulated."[11]

To be sure, law—and the coercive power of the state to enforce law—tends to be the mechanism invoked by society to shape behavior only after all other avenues have been exhausted and found not to be effective. These other avenues include friendly persuasion, parental admonition, social pressure, contracts, licenses, or informal agreements. How such costs and benefits are weighed by those involved is an important factor in determining what avenue is more appropriate for the given problem to be solved. Through these other avenues, codes and standards of acceptable behavior and etiquette are established and enforced. Indeed, this is one key purpose of education: to socialize people into internalizing the rules of behavior that the community has come to accept as reasonable.

On electronic networks, all of these nonlegal mechanisms are used. For example, a popular book called *Zen and the Art of the Internet*[12] describes a set of "netiquette" rules—rules that Internet users are inclined to accept and expected to follow. Newcomers who behave in a manner inconsistent with these rules are subject to reprimand from the group.

With most commercial network service providers, a condition of use to which all users must agree is that users will abide by a certain set of rules about acceptable behavior; violators can be punished by the provider's discontinuing their access to the system. Contractual

---

[11]On the other hand, a legal regime does exist for space today, as does one for the sea. Both are aspects of international law. International dimensions of network connectivity will increase as more nations connect to a global network, but the extent to which international law will be relevant remains open. The reason is that in the absence of a supra-national mechanism to enforce international laws, such laws are enforceable only with the acquiescence of a violating party. For example, a treaty to which the United States and the former Soviet Union are parties forbids the placement of nuclear weapons on the moon or on the terrestrial seabed. The parties refrain from taking such actions because they do not regard the taking of such actions as being in their long-term self-interest. However, an intentional and continuing violation of this treaty by one party could result only in unilateral actions taken by the other party, since there is no supra-national mechanism to force compliance with the treaty. In general, what keeps parties in compliance with treaty obligations is that the parties have more to gain in the long run than they have to lose, rather than the fact of any external enforcement of such compliance. But as parties that do not play by such rules connect to the global network, increasing pressures for more effective enforcement mechanisms may result.

[12]Brendan P. Kehoe, *Zen and the Art of the Internet: A Beginner's Guide*, 2nd Ed., PTR Prentice Hall, Englewood Cliffs, N.J., 1993.

relationships between users and providers can be tailored according to the needs of the parties involved, and they often reduce the delays of legislation and the costs of litigation. Others question this strategy, arguing that the contractual relationship is not an even playing field because the user is most often a single person rather than a corporation, and the provider is usually a corporate entity with financial resources far superior to those of most individuals.

Bulletin board operators often operate in a similar fashion: they write their own rules and set them forth the first time a new user logs in. Enforcement provisions are similar to those used by commercial providers. The playing field is arguably more even because most bulletin board operators are relatively small in scale, and thus potential individual users have relatively more clout and greater freedom of choice.[13]

Many universities have codes of acceptable network behavior as well. For example, EDUCOM has promulgated a bill of rights and responsibilities for users in the academic community that it urges its members to respect. When adopted, such codes are a part of the rules and regulations governing the academic community, and all relevant academic mechanisms for enforcing rules and regulations are available to university policymakers for dealing with miscreant electronic behavior.

Kiesler suggested that human behavior, both good and bad, is more extreme on electronic networks than in other arenas, thus underscoring the need for effective ways to regulate behavior. "I don't mean that people turn into terrible beasts on networks, but on average there is more misbehavior, there is more gossip, there is more 'flaming,'[14] there are more extreme opinions, and there is more harassment on networks than there is in other areas," she said. Moreover, the consequences may be worse on networks than in other situations, in that more people are affected and false rumors can spread beyond a local population, Kiesler suggested. She added, "The issue is not how we crack down on people, but how we promote more prosocial behavior. . . . What do we do to promote more responsibility taking [and] ethical and moral behavior in network communi-

---

[13]The importance of a user's ability to choose among various service providers is underscored by the administration's position regarding user responsibilities: users rather than government or providers should be the ones to protect user sensibilities.

[14]"Flaming" refers to the making of insulting and ad hominem attacks against another person, generally without much reflection or thought but with a great deal of (presumed) emotion.

ties?" How this will evolve in the future is an open question, and speculation on whether civility and decorum and respect for others will increase (as more users demand standards for reasonable behavior) or decrease (as the presence of more users leads to greater fragmentation on the network) is just that—speculation—at this time.

John Perry Barlow, a co-founder of the Electronic Frontier Foundation, framed the issue as part of the challenge of community development. Cyberspace is the community of the future, but it is still emerging, Barlow said; it is "the latest thing in suburbs" as opposed to an updated model of the cohesive small town. Cyberspace still lacks a sense of community experience, which provides a seasoned benchmark for evaluating information, and a sense of responsibility, which derives from development of individual identity, he said. Barlow argued that it was necessary for society to trust a messy process of sorting out community ethics, values, and conscience rather than turn to the law, which by its nature cannot be expected to provide adequate guides to behavior in a rapidly evolving environment. He further suggested that a sense of personal responsibility has been difficult to develop or enforce on electronic networks because of conflicting desires to both authenticate and conceal identities.

Governance of electronic network communities poses special challenges and opportunities. Using a "many-to-many" communications mode, all members of an electronic community can in principle be heard and their comments made part of the official record (this is the root of the political electronic marketplace metaphor described in Chapter 1). This raises the interesting question of the role of elected representatives and how they are to participate in the system. In electronic communities, peer participation and consensus building through iterative discussion are often the rule, but the active participants jealously guard their rights to determine for themselves the rules under which they will operate.[15] Some commentators have

---

[15]A major complication to the very concept of consensus—present in most nonelectronic communities but particularly important in electronic ones—is that most electronic groups include many individuals who do not make active contributions but instead are content to remain "lurkers" who just listen. Indeed, in most groups, the number of people encompassed by a mailing list (i.e., the number of people who receive messages addressed to that mailing list) is much larger than the number of people who ever send anything in. Thus, "consensus" can be achieved only among the people who are willing to engage in a discussion.

even suggested that electronic networks are fundamentally a medium conducive to democracy.[16]

Although many speakers at the workshop and forum argued that community values and ethics should be given priority, other speakers emphasized that the ethical foundation for behavior on electronic networks was shaky, due largely to a lack of consensus about what constitutes "reasonable" behavior. For example, Branscomb said that where electronic networks are concerned, ethics and the law often conflict, due to shortcomings in the latter. She observed that "when the laws don't make [sense] it erodes ethical values, and that is what is happening, to a large extent. I know a lot of the computer scientists have said the laws really are a pain, they don't make any sense. . . . I think a lot of that is happening with the copyright law [for example]. It's easy to replicate things, and the copyright rules don't make sense to a lot of us [the users], so we just ignore them. And therefore, we really are in the process of developing the ethical values which we have to impose upon the system before we can decide what laws we want to enact."[17]

Nonetheless, once laws are passed, all those within the applicable political jurisdiction are subject to those laws, and so the existing legal regime does affect networks; existing law does affect and constrain the behavior of all of us. Instead of asking whether or not law affects behavior on networks, a more useful question might be, How should laws be formulated so that they produce the desired effect? Speaking of another medium, Westin attributed the success of certain laws to their origins in the community that those laws would affect. He specifically cited the privacy provisions in the Cable Communications Policy Act of 1984 (Public Law 98-549), which were based on a code developed years before by Warner Amex. He further suggested that laws regulating behavior on electronic networks might well and fruitfully be based on the expectations of reasonable

---

[16]David Hughes argued that the founding fathers would have used computer bulletin boards to correspond. This may be true; Benjamin Franklin was the originator of the system of surface transport of written information called the U.S. Postal Service, what we now call "snail mail"—the only technology readily available at the time for exchanging the ideas upon which this nation's system of governance is built.

[17]It is worth noting that ethical values are only part of the issue. Changing economic balances also matter greatly—easier replication also means more replication, and increases in replication raise its potential economic significance. The economic point is discussed further in Computer Science and Telecommunications Board, National Research Council, *Realizing the Information Future: The Internet and Beyond* (National Academy Press, Washington, D.C., 1994), p. 162.

behavior developed by the community in question rather than imposed as pronouncements from outside.

A question argued repeatedly in both the forum and the workshop was the extent to which special legislation is needed to govern the new networked communities. According to this view, there is something new and unusual about the new electronic environment that may need special legislation to protect its unique qualities. Others argue that there is no particular need for network-specific laws, as existing laws for the most part cover miscreant behavior that might occur on networks. Moreover, to be effective new legislation requires a social consensus both about what is new or unique about networks as compared to other media and about what values that legislation will embody. Such a consensus does not exist today.

3

Legal Considerations for Electronic Networks

Many users and providers of network services are wary of lawyers' influence on networked communities. Jack Rickard believes that "when the law appears, it's death upon contact." His plea, shared by most of the bulletin board system operators (known as "sysops") is: "Just leave us alone." At the same time, advocates of this view do recognize the need for defining socially acceptable behavior; their method of choice for promoting such behavior is education that appropriately socializes participants in networked communities.

Nonetheless, existing law does have an impact. Law is important because when something goes wrong, those harmed consult lawyers for help. To determine what they can do to obtain redress for such harm as may have been done, lawyers look to existing law for precedent. We are a very litigious society, and so the courts have a vast amount of experience in sorting out our demands for justice. Judges themselves turn to existing law to determine how to seek fair and equitable solutions to our problems. Policy analysts use the existing precedents to see what works and what does not. Finally, ordinary people rely on existing law because users carry their expectations with them from one environment to another.[1]

NOTE: In this chapter, all quoted material that is not otherwise identified originated with the individual noted, speaking at the November 1992 workshop.

[1]Sara Kiesler points out that the officers charged in the Rodney King beating used their electronic communications system as though it were a private telephone line, even though they had been warned that all traffic over that system was recorded. These conversations were later introduced in court as evidence against them.

LEGAL DOMAINS

Existing legal domains (Box 3.1) offer a rich heritage. The fundamental rights contained in the U.S. Constitution and the Bill of Rights are a good beginning. Included are the copyright and patent systems, prohibitions against denial of human rights without due process, the right of free speech and assembly, immunity from unreasonable search and seizure, and protection against self-incrimination. A long history of Supreme Court decisions finds that the Constitution contains an implied right of privacy, although the boundaries of this right are subject to considerable debate. The Constitution is especially important in protecting the rights of minorities against the majority rule.

Common law tends to develop slowly and deliberately, based on common sense and logic and the customary expectations of the community. It is conservative, stable, and, some fear, too rigid, but it does provide a workable framework for analysis, even though the frame does not always fit exactly. Thus, despite rapid technological change, electronic networks do not constitute a lawless frontier in which individuals may do anything they please.

The long history of common law complements the basic constitutional foundations by providing alternative strategies for protecting property interests and obtaining redress for grievances for damage occasioned by negligence or willful disregard of harmful consequences that might ensue. In other words, the law expects human beings to behave in a somewhat rational and respectful manner in their dealings with each other and with regard to their proprietary interests. Case histories in property law, tort law, and equity provide ample precedents to guide the behavior that might be expected to be reasonable in an electronic environment.

There are also numerous explicit statutes covering computer use and misuse (all but one state has enacted such a statute). Statutes also cover privacy but are usually specific to certain kinds of information, such as credit histories and medical data that are particularly sensitive. The Electronic Communications Privacy Act of 1986 specifically forbids eavesdropping on data traffic. Lawyers and their clients also must be aware of the laws concerning commercial transactions, including antitrust provisions.

Several regulatory agencies have administrative jurisdiction over certain types of network activity. The Federal Communications Commission (FCC) has authority to license common carriers and is active in promulgating standards for high-definition television transmission and telephone network architecture; it also regulates cable television and certain uses of the electromagnetic spectrum. The Federal Trade

Box 3.1
Legal Domains

Domain	Examples of Domain	Sources
Constitution	Free speech Copyright/patents Due process Equal rights Freedom from search and seizure	Congress/President U.S. Supreme Court
Common law	Tort Equity Property Intellectual property Contracts	Courts
Statute	Electronic privacy Computer crime Contract Antitrust	Congress and state legislatures, popular initiatives and referenda Executive branch action to implement legislation
Regulation	Code of Federal Regulations (e.g., approvals required by Section 214) Open networks	Administrative agencies (Federal Communica- tions Commission, Federal Trade Commission, and others) Judicial review Legislative oversight

NOTE: In all instances, the enforcement mechanisms consist of monetary damages, specific enforcement of contract provisions, incarceration, and/or fines.

Commission has authority to regulate the content of advertising so that it does not mislead consumers. The National Institute of Standards and Technology plays a major role in the formulation of Federal Information Processing Standards. State public utility commissions have overlapping jurisdiction over intrastate network traffic, thereby confounding, confusing, and complicating the legal environment within which network providers and their users operate.

Finally, judicial interpretation and oversight affect all legal frameworks and regimes. Since it is often unclear how existing statutory and/or case law ought to apply to any given instance involving electronic networks, the views of judges and juries will often set new precedents in this arena. Moreover, since in practice it is often expensive to litigate in defense of one's position, unfair or inappropriate legal advantages may accrue to those who have the resources to defend a potentially controversial interpretation of existing law. In the absence of legal challenge, such interpretations may themselves become new precedents.

LEGAL MODELS EXISTING IN THE ELECTRONIC ENVIRONMENT

Legal models can provide useful analogies that help to sift out what is different from what is similar. For the purpose of analysis, consider the different legal models listed in Box 3.2 that exist within the network environment.

Perhaps the most important legal models for the network environment are those of publisher and distributor, primarily because of the question of liability. Publishers generally are held to a higher standard of liability for content than are distributors, because publishers have the capability to (1) control access to the medium by others seeking to disseminate their own information, (2) review and gain knowledge of the content prior to publication, (3) alter or exclude content prior to dissemination, and (4) require attribution or permit anonymity or confidentiality for content.[2] A firm that exerts

---

[2]For some analysts, the distinction between capability and the actual exercise of those capacities may be an important issue in defining a concept, and responsibility and liability for content, in the electronic environment, since a duty to act does not depend on whether action is ultimately taken but merely on whether it should have been taken. For example, Allan Adler asserted that ". . . what I mean by capabilities is, if you have the ability, technologically and physically in real time, [and] the authority and opportunity to do these things, you may be legally considered a publisher rather than a distributor of information, with the appropriate level of legal liability." (See the discussion of the *Cubby* decision in Chapter 4.)

Box 3.2
Legal Models to Characterize a Network Environment

- Publisher—Prodigy Services Company (for many services)

- Distributor—CompuServe

- Cooperative—EduNet, NEARnet

- Library (information provider)—LEXIS, DIALOG, Medlars

- Private networks

    – Corporate—IBM, Hewlett Packard, Citicorp, and others
    – Personal—Bulletin board system operators ("sysops")

- Common carrier—MCI, AT&T, Sprint, regional Bell holding companies

- Mixed or hybrid—cable television

- Trusteeship—broadcasters

- Marketplace—information entrepreneurs marketing their wares

- Information utility—Santa Monica Public Electronic Network and Cleveland FreeNet

---

editorial control over the electronic traffic it carries (as does the Prodigy Services Company over much of the information content it delivers on-line regarding sports, news, features, and so on) may claim the legal status of a publisher, but in so doing it also subjects itself to the liability thereby implied. A firm that does not exercise direct editorial control over the traffic it carries (e.g., one that subcontracts the management of editorial content to a third party) is arguably a distributor, with a correspondingly lower level of liability.

Next in importance as a legal model for the network environment is the cooperative—an arrangement whereby independent entities of equal status come together to achieve a common purpose and to share costs as well as profits. EduNet is best classified as a cooperative because it is put together so as to have some legal weight. Certain regional networks such as NEARnet and NYSERNet also operate as cooperatives. The Internet as a more collaborative effort does not qualify legally as a cooperative, although various networks interconnected through the Internet have a cooperative arrangement that may be evolving toward a cooperative as defined legally.

Also important, if only because it is so often ignored, is the information utility. (Here the term "utility" is used in the sense of a service that must be provided to essentially anyone in the relevant community or service area. It is analogous to the concept of "common carrier" as used in the world of telephony.) Only a few publicly funded networks, such as the Santa Monica Public Electronic Network (PEN) and the Cleveland FreeNet, are true information utilities. Many commercial services look like information utilities in that they offer subscribers a potpourri of information services. However, such services do not represent themselves as information utilities, since doing so would require them to provide access and some minimum level of service to all potential users. Government-operated information networks that operate as public forums are a different matter; under the First Amendment, government-operated networks do not have the right to discriminate among users or to monitor or control messages and thus must provide access to everyone in the community. Distinctions among government-operated networks, government-sponsored or government-funded network services, and public resources offered over privately operated networks raise yet other questions for access that are as yet unsettled.

It is instructive to identify the various rights and responsibilities of different types of participants in the network environment (Box 3.3). As a general rule, it seems to be the case that the more an entity undertakes to do or the more value-added services it undertakes to provide, the higher the degree of liability for which it is responsible. The creators of information are provided legal protection (and restraints) through the copyright and patent laws. Publishers are protected primarily under the First Amendment. Distributors govern their relationships with their sources and their customers through contract; a customer dissatisfied with the services provided has one primary option—to take his or her business elsewhere. Common carriers are subject to an elaborate regulatory system established by law and administered by the FCC and state regulatory agencies and commissions. At present users are governed largely by the "netiquette" they have established by custom, contract, or common law if they want to litigate about some harm that has occurred. Broadcasters (radio and television) carry the burden of trusteeship; that is, they are licensed to act as trustees of a public property (the public airwaves) and have a legal responsibility to act "in the public interest."

Box 3.3
Rights and Responsibilities Distinguished by Function

Function	Rights	Responsibilities
Creator	Control of content Compensation Integrity	Originality Liability for damage
Publisher	Control of content Compensation Integrity	Liability for damage
Distributor	No control of content Compensation	No liability for damage
Carrier	Limited liability for transport Compensation	Fidelity of carriage Integrity Timely delivery Provision of equitable access

User	Accessibility
	Equity
	Due process
		Avoidance of:
		  negligence,
		  abuse,
		  misuse, and
		  misappropriation

Trustee	Licensee of public property	Action in the public interest

WHAT IS DIFFERENT ABOUT ELECTRONIC NETWORKS?

To the extent that electronic networks are a new medium for discourse and communication, new laws or new interpretations of existing laws may be necessary. Thus, it is necessary to identify how electronic networks might be "new" from a legal perspective.

One obvious difference is that territorial jurisdictions are largely irrelevant to the electronic environment,[3] except perhaps in the ini-

---

[3]This is true in the sense that once connected to a network, a user can often electronically cross political boundaries with relative ease. However, territorial jurisdictions may determine who has access to a given network (e.g., a city may grant access to a public network only to its residents; a university may grant access only to students on campus).

tial stages when spectrum must be allocated for the purpose where wireless communications are employed and licensing authority must be obtained unless regulatory authority is waived. For example, in the United States, information providers that offer users services that go beyond "basic" services have been permitted to flourish with minimal government interference.[4]

More confusing, perhaps, is the fact that many legal precedents have been established in the context of a given technology—e.g., broadcast or cable, telephone, or mail. In considering the legal implications of a new medium, analysts quite naturally look to those precedents for guidance. But it is never certain how old precedents should apply to new media, and so neither analysts nor users of a new medium are certain which precedents should guide their analysis or govern their use.

As noted in Appendix A, e-mail can be used to support a mode of communications often known as bulletin boards in which a message posted to a bulletin board by one user is distributed via e-mail to all of the other subscribers to that bulletin board. Some electronic bulletin boards start as a mechanism to deliver private communications among a small group of friends. To the extent that the term "bulletin board" makes one think of a little bulletin board on one's refrigerator, the traffic on the electronic bulletin board is clearly outside the reach of the law. But electronic bulletin boards are also a way of communicating with large numbers of people in many different geographic locations (perhaps with differing laws regarding message traffic).

The very term "e-mail" makes one think of the electronic analog to a postal letter; however, e-mail does not, as a matter of law, enjoy the same status as a postal letter. Entirely different legal precedents and principles may in fact apply. For example, some users of e-mail have the same expectations of privacy for e-mail that they do for postal mail—an issue that has generated considerable legal controversy.[5] Users of private e-mail systems may expect their messages

---

[4]At the same time, the definition of what counts as "basic services" is a matter subject to considerable debate. For many years, "touch-tone" service provided by the local telephone company was regarded as an "extra" not included in the basic service definition. However, touch-tone service has increasingly become a prerequisite to navigating through other services provided by telephone (e.g., voice mail), leading to pressures to treat it as part of "basic" telephone service.

[5]For example, in a 1993 decision, a California appellate court found in favor of an employer that had terminated an employee after reviewing his e-mail messages and finding among them personal messages of a sexual nature to other employees. Specif-

to be private, and indeed they are protected in that expectation to a certain extent by federal statute (e.g., public officials may not tap those communications without judicial authorization); but this expectation may not be appropriate in a corporate e-mail environment where the corporation expects to manage the traffic for business purposes and in its corporate interests.[6] The privacy principles drafted by the Information Policy Committee of the Clinton administration's Information Infrastructure Task Force may ultimately affect these expectations.

A further complication is that it is not only electronic networks that are assuming new functions: established media are also doing so, and there is a lack of consensus on which models and precedents are appropriate for what purposes. The regional Bell holding companies[7] (RBHCs) want to be information service providers as well as carriers. Electronic bulletin boards are becoming networks that allow messages to pass through them unread; a good example is Fidonet. Newspapers are distributors as well as publishers, as the advertising supplements supplied by advertisers indicate. Broadcasters are offering conferencing facilities in electronic town meetings, and both radio and television (using telephonic call-ins) have turned into very interactive media on the talk shows. Cable television systems, traditionally delivering entertainment programming to the home, are geared up to provide data services to businesses and to experiment with home-based information services. Complications are exacerbated by the fact that different providers may have been regulated under different assumptions (as is the case for the RBHCs, long-distance telephone service providers, and cable companies).

---

ically, the court held that the employee had no reasonable expectation of privacy since he had signed a form acknowledging a company policy that restricted employees' use of computers to company business. In a second court case, still under litigation in January 1994, a company sued a former employee on the basis of messages retrieved from the former employee's electronic mailbox (which was not maintained or provided by the company). A key issue will be whether the company has the right to access the electronic messages of its employees. (See Michael Traynor, "Computer E-Mail Privacy Issues Unresolved," *The National Law Journal*, January 31, 1994, S-2.)

[6]Many corporations do take the position that e-mail messages are subject to review in the same way that hard-copy letters and memoranda are subject to review, and they do post notices periodically (e.g., every time the user logs into the host system) to this effect. However, as Sara Kiesler pointed out at the forum, people habituate to such warnings very rapidly, and the net effect is that they respond as though those warnings were never given at all.

[7]The regional Bell holding companies are Nynex, Bell Atlantic, US West, BellSouth, Ameritech, Southwestern Bell, and Pacific Telesis.

The most important difference, and one very troubling for existing law, is the easy ability to replicate, alter, and manipulate images and text. This ability changes established legal concepts—e.g., the notion of when something is published. Indeed, by accessing previously developed information, anyone with a computer can become a creator of an information product or the producer of a library or database that distributes or acquires information assets with financial value. An example is the provider of a value-added service that searches public records or documents and makes them accessible, for a fee, to users in a much more convenient form; proprietary rights have been asserted for such products, even when the underlying information is public and freely available. Today's legal regime provides mixed signals and expectations concerning what is public-domain information and what is proprietary information entered into the electronic environment. There is a need for more clarity concerning the nature of the traffic and its legal status. In presenting this capability to manipulate and disseminate information, electronic networks provide the underpinning of an electronic marketplace in which some information is shared, some is sold, and some is merely transmitted.

The challenge today is to devise a legal regime appropriate for the capabilities enabled by the technology now available. This regime necessarily includes certain aspects of existing law (and precedents and interpretations applicable to network technology) and may include new laws specifically needed for this technology.

LEGAL CONSIDERATIONS THAT AFFECT
THE USE OF INFORMATION

A number of legal considerations pertain to protection of the rights of network users.

- *Secrecy* is designed to prevent deliberate disclosures of information to individuals not authorized to receive that information. The information being kept secret may refer to the contents of a message (a consideration among lawyers known as security) or to the identity of a message's author (a consideration known as anonymity). The inverse of anonymity is known as authentication, or the assurance that the putative originator of a message is in fact the real originator. Closely related is the notion of nonrepudiation, the property that the putative originator of a properly signed message cannot plausibly deny his or her originating that message.
- *Privacy* is intended to prevent intrusions into information about

individuals by persons not authorized to have access to that information. Privacy law today is ambiguous; it is used to cover many different concerns about misuse of information. To some, privacy should be more strictly construed to mean the right to prevent unwelcome and unauthorized intrusions such as obnoxious telephone calls, junk mail, and overeager journalistic inquiry—in other words, any entry into the personal space that should be private. Such a view is reinforced by an oft-forgotten Supreme Court decision that confirms that one's home is one's castle.

Both privacy and secrecy are mechanisms that have been invoked to reduce the accountability of individuals under the law.

* *Confidentiality* is the right to release information with restrictions on its redistribution. There are a few common-law precedents that recognize personal confidentiality: the right of a spouse to not testify against a mate, the sanctity of the confession to one's religious mentor, and the attorney-client privilege are the most obvious. However, most of the areas that have come to be considered confidential have developed by explicit statute achieved through the efforts of concerned citizens. Today there is a considerable effort to achieve a more uniform way of treating confidentiality of medical records, in order to assure individuals a measure of personal privacy and to not inhibit too strictly the transfer of medical information that would benefit both the patient and the public's interest in a healthier society.

* A *publicity* right is one that gives each individual the right to disclose information freely in the public domain or to demand compensation for public release. The abandon with which many participants on radio and television talk shows disclose very personal information about their own behavior and inner thoughts confirms that this is a right alive and well today. There is also no dearth of material generated by public relations firms for all manner of public figures who seek media coverage. However, what is missing is a clearly delineated point that defines when something is intended to become public-domain information, when it is intended to remain confidential between the communicants on electronic networks, and when it is released with the expectation of remuneration for redistribution.

* *Commonality* remains the right of the body politic to determine what areas shall remain off limits to proprietary appropriation. Laws of nature and languages are among the few areas clearly beyond proprietary appropriation. The controversial area at the moment is in the area of "facts," which are usually assumed not to be

subject to copyright protection because they are not original expression. However, we have not yet clarified what nature of "facts" might appropriately belong within the protected area of personal privacy and which are more likely to be required to be disclosed because of a higher public interest.

- *Liability* is the domain of tort law where acts of negligence, abuse, neglect, and misuse cause damage that can be redressed through litigation in the civil courts. (*Civil liability* is often synonymous.)

- *Criminality* is the domain of the penal statutes, which impose more stringent sanctions involving fines as well as incarceration, or terms of probation during which behavior is carefully monitored. (Criminal liability is a related concept pertaining to the individual who may violate a penal statute.)

- *Accessibility* involves a right to participate and to reply. As a rule, the First Amendment gives information providers the right to determine the content of their own publications. However, this right is not unlimited, and under certain circumstances, information providers may be compelled to provide content that they might prefer not to carry. Examples include the fairness doctrine of the FCC, under which broadcasters (e.g., radio and television stations) are required to provide air time for those representing views that oppose those expressed by those stations; laws that carve out specific exceptions to First Amendment rights (e.g., laws banning cigarette advertising in certain media); and regulatory policies intended to guarantee free and open political debate. As a consequence, access is a right more often sought than exercised in practice.

- *Accuracy* is a right to correct information inaccurately recorded. With the advent of computerized records that can contain a vast number of similar names, opportunities continue to increase for inaccuracies to proliferate and cause embarrassment or harm, whereas the law governing rights to obtain access to records and to justify corrections is not well developed.

- *Integrity* is the right to protect an original work according to the desires and wishes of the creator, a right that has not been protected in the United States so much as in Europe, where "moral rights" of authors and artists have been recognized. The recent effort of some in the motion picture industry to protect the integrity of black-and-white film classics is a good example of an effort to ensure the integrity of information. What the proponents of "integrity" achieved was the National Historic Film Preservation Board, which could designate films as "classic" with the proviso that they be so labeled in their original form. However, the Register of Copyrights nonetheless authorized colorized movies of these classics to be marketed under a separate copyright issued to protect the colorization as a new expression of the work. In an electronic bit stream, where even a few pixels can be plucked out of a work, it is going to be difficult to determine what rights of integrity can be protected in the original work.

• *Equity* is an important legal theory in common law that no wrong must go unrighted. If harm occurs, then society must find a way to redress it. When lawyers can find no statute, precedent, or commonly accepted custom, they fall back on general principles of equity.

Finally, *interoperability* is a property of computer systems that facilitates the use of hardware and software of different models and vendors in a compatible manner. Although interoperability does not qualify as a legal consideration per se (if it did, the problem of incompatibility would be a phenomenon of the past), a variety of court decisions (some discussed below) have marked relevance to it; in addition, as the National Information Infrastructure moves from concept to reality, evolving and assuming increasing prominence, a number of legislative and regulatory factors may begin to influence interoperability.[8]

COURT CASES LAWYERS RELY ON FOR AUGUMENT

Lawyers and their clients look to existing case law for guidance in arguing their cases before the courts. They can choose from hundreds of court cases in which the established communications media have litigated their differences and grievances. Box 3.4 summarizes several of the most important cases.

The *Cubby* case (Box 3.5) is the first case to be decided with respect to commercial network providers. In essence, the finding in *Cubby v. CompuServe* was that a party acting as a distributor only—and that thus cannot monitor the traffic and consequently cannot know or have reason to know of the content—has no liability for offending remarks in the content. It is not clear that the judges understood that the technicians could actually monitor the traffic, but the decision does offer some protection.

However, *Cubby* may be a more dangerous than helpful case because it could lead many providers of network services to assume

[8]For more discussion of this point, see Computer Science and Telecommunications Board, National Research Council, *Realizing the Information Future: The Internet and Beyond*, National Academy Press, Washington, D.C., 1994.

44   RIGHTS AND RESPONSIBILITIES IN NETWORKED COMMUNITIES

Box 3.4
Some Relevant Legal Precedents

Case	Result
Cubby Inc. v. CompuServe	No obligation to monitor
Feist Publications Inc. v. Rural Telephone Co. Inc.	Facts not copyrightable
Sega Enterprises v. Accolade Inc.	Reverse engineering for interoperability is permissible
Armstrong v. Executive Offices of the President	Electronic files are public records
Playboy Enterprises Inc. v. George Frena et al.	Intent to infringe on copyright is not required for a finding of infringement
Ford Motor Credit Co. v. Swarens	"Trust in the infallibility of a computer is hardly a defense. . . ."

that they can rely on an assumption that they are distributors of information only, with no responsibility for the content of their offerings. Such service providers may assume that they reside in the same legal niche as the bookstore or the newspaper stand, which may not be the right niche or the ultimate niche into which the law will assign these new electronic services. One danger could come from trying to apply the publisher model to electronic bulletin boards. Surely there is some area in the electronic environment that should rightly be considered a private space. This would be comparable to a private home or club where friends and peers may share private and confidential communications. Is posting to a controlled-access bulletin board "publication"? And how do we characterize public conferences to which users contribute their comments, often with no centralized editorial control or managing supervisor?

The application of a general law across the entire electronic community may well result in curtailing some activities that really ought to be permitted to continue. Legal precedents such as *Cubby*—or,

---

Box 3.5
Cubby v. CompuServe, 1991

CompuServe is an on-line information service that grants access to a variety of information sources in exchange for a fee. These information sources are provided for CompuServe contractually; that is, CompuServe contracts with vendors to provide information, but then makes this information available to users through a mechanism called forums. The provider of one information source, Cameron Communications Inc., agreed in its contract with CompuServe to "manage, review, create, delete, edit, and otherwise control the contents" of its forum.

Defamatory statements about a firm known as Cubby Inc. appeared in the forum operated by Cameron Communications. Cubby Inc. sued CompuServe for libel, based on the defamatory statements. CompuServe moved for summary dismissal on the grounds that it acted as a distributor, not a publisher, of those statements and thus was not liable for the statements.

In 1991, the U.S. District Court in the Southern District of New York granted CompuServe's motion for summary dismissal on the grounds that CompuServe was indeed acting as a distributor and indeed had little or no editorial control over the contents of a contracted information source. The court further reasoned that CompuServe had no reasonable opportunity to review for defamatory material every information source it made available.

SOURCE: *Cubby Inc. v. CompuServe*, 776 F. Supp. 135 (S.D.N.Y. 1991).

---

more accurately, the implications of their distinctions—may come to be applied too broadly.

The *Feist* case (Box 3.6) clarified the outer boundaries of the copyright law, holding explicitly that facts are not copyrightable. The extent of labor expended—the "sweat of the brow"—is not to be considered in granting protection. This case will likely create turmoil in the law because it is facts, now, that people are most concerned about—e.g., telephone numbers, names and addresses, and transaction-generated information (data about what one purchases or how one travels). The current legal regime does not recognize a right to what might be called personal autonomy over information about oneself, and without copyright protections on compilations of these facts, such compilations may be disseminated and reproduced quite widely.

The *Sega* case (Box 3.7) involves a controversial area with respect

---

Box 3.6
Feist v. Rural Telephone, **1991**

Pursuant to state regulation, the Rural Telephone Company Inc. publishes a "white pages" telephone directory that lists telephone subscribers alphabetically and their corresponding telephone numbers for a certain geographical area of service. In order to produce a similar "white pages" directory for a much larger geographical area that included the service area in question, the Feist Publication Company extracted the names and numbers it needed from the Rural Telephone Company's white pages without Rural's consent. Although Feist altered many of the listings it obtained, several were identical to those listed in Rural's white pages. Rural sued Feist for copyright infringement.

In 1991, the U.S. Supreme Court ruled that Rural's white pages were not entitled to copyright, and thus that Feist was not liable for infringement. The basis for this ruling was that the selection, coordination, and arrangement of Rural's white pages did not satisfy the minimum constitutional standards for copyright protection, since Rural had only to obtain data from its subscribers and list them alphabetically, resulting in a "garden-variety" white pages directory devoid of even the slightest trace of creativity (which copyright law is intended to protect).

SOURCE: *Feist Publications Inc. v. Rural Telephone Co. Inc.,* 111 S. Ct. 1282 (1991).

---

to fair use of computer software, an area of the law that has not yet been reviewed by the Supreme Court. It concerns the question of what amount of reverse engineering of software is permissible in order to understand the underlying concepts, which are not protected by copyright law, and knowledge of which, according to an amicus brief prepared by a group of intellectual property academics, is essential to achieve interoperability of software.[9] Incompatibility of software (lack of interoperability) is a major deterrent to the rapid development of networks and an area of the law demanding of public attention.

What the academics seem to be saying is that there is a limit to what the copyright law can cover—that copyright law offers very

---

[9]Some additional discussion of these issues is provided in Computer Science and Telecommunications Board, National Research Council, *Intellectual Property Issues in Software*, National Academy Press, Washington D.C., 1991.

Box 3.7
Sega v. Accolade, 1992

In order to develop video-game cartridges that were compatible with Sega Enterprises' "Genesis" entertainment system, Accolade Inc. disassembled the computer programs contained in Sega's video-game components to learn the requirements for compatibility. Accolade then produced a manual that described the functional requirements of compatibility with Sega products without including any of Sega's original computer programs. Based only on the manual, Accolade then wrote its own programs that were compatible with the Sega interface. Sega sued Accolade, charging copyright infringement.

In 1992, the Ninth Circuit Court of Appeals overturned a lower court ruling that Accolade had infringed on Sega's copyright of its computer programs. In particular, the Court of Appeals held that Accolade's use of the "disassembled" programs was protected under the fair-use provisions of the Copyright Act if disassembly is the only way to gain access to the ideas and functional elements embodied in a copyrighted computer program.

SOURCE: *Sega Enterprises v. Accolade Inc.,* U.S. Court of Appeals, Ninth Circuit, No. 92-15655; D.C. No. CV-91-3871BAC, October 20, 1992.

---

thin protection for computer software interfaces, in fact protecting only the expression that is different from all others and proprietary to the creator. It may be that this case marks a turning point that indicates the limits of the judicial system in carving out new protections for computer software. Users and providers alike may now turn to Congress to sort out what is optimum for the software industry and network users nationwide. In the European Community the policy with respect to reverse engineering and fair use of proprietary interfaces has been hammered out in an open and public discussion, not in the courts.

The *Armstrong* case (Box 3.8) held that electronic communication can be a public record and be subject to the Freedom of Information Act, the Federal Records Act, and the Presidential Records Act. A case with similar elements was raised in May 1994 that involves the refusal of a state university to disclose e-mail messages sent to university administrators.[10] The individual wishing to see those mes-

---

[10]See Thomas DeLoughry, "University of Michigan Refuses to Release E-Mail of Administrators," *Chronicle of Higher Education,* Vol. 21, p. A-28, January 26, 1994.

Box 3.8
Armstrong v. Executive Offices of the President, 1993

The U.S. Court of Appeals upheld a lower-court ruling that the Executive Offices of the President must preserve electronic documents as "federal records" under the Federal Records Act, affirming that hard-copy printouts were insufficient because they might omit "fundamental pieces of information which are an integral part of the original electronic records, such as the identity of the sender and/or recipient and the time of the receipt."

SOURCE: *Armstrong v. Executive Offices of the President,* 1 F.3d 1274, 1993, U.S. App. LEXIS 20527 (D.C.Cir. 1993).

---

sages contends that they are analogous to written letters and thus subject to disclosure under that state's freedom-of-information law. The university contends that e-mail messages are analogous to telephone calls, which are not subject to such disclosure, and in addition that they are protected under the federal Electronic Communications Privacy Act.

Both cases raise the issue of whether electronic communications in the form of e-mail constitute "records" that are subject to the intent of the various freedom-of-information acts. Such cases also raise the issue of what types of electronic communication should be kept. For example, should all e-mail used by federal agencies be kept as public records? In the days when the federal government operated by telephone calls and written documents, it was clear that telephone calls did not constitute public records, though written notes about those calls might. Since many people use e-mail instead of a telephone for reasons of convenience (e.g., avoiding telephone tag), some argue that at least some e-mail should be treated as a telephone call would be. On the other hand, requiring the sender of an e-mail message to distinguish between a noteworthy and non-noteworthy note presents problems as well, especially if that determination must be made at the time of transmittal.

The *Playboy Enterprises* case (Box 3.9) was decided recently. In this case, the court placed on bulletin board operators that charge fees for access a substantial burden to review files accessible on their systems for possible copyright violations, though the extent of this burden was not definitively established by this case.

Finally, an early but important computer case is *Ford Motor Credit*

Box 3.9
Playboy Enterprises v. George Frena, 1993

The *Playboy Enterprises v. George Frena* case involved a large number of images identified as pictures taken from *Playboy* magazine and protected by copyright that were found in the form of graphics files on a bulletin board system owned and operated by George Frena; these files were not posted by Frena, but rather by users of his bulletin board service. The bulletin board charged fees for users to access these (and other) files. The court held that this use was clearly commercial and that the bulletin board operator, George Frena, was guilty of copyright infringement. Though Frena argued that he had no knowledge of the source of the images, the court held that intent to infringe was not essential to a finding of infringement.

SOURCE: *Playboy Enterprises Inc. v. George Frena et al.,* 839 F. Supp. 1552, U.S. Dist. Ct., M.D. Fla. Dec. 9, 1993.

Company v. Swarens (Box 3.10). An old maxim proclaims that ignorance of the law is no excuse. This case held, in a similar vein, that one cannot rely on faith that the computer will be accurate as a legal defense. Put differently, the computer does not enjoy special status as an infallible agent, especially when it can be shown not to be infallible.

SPECIAL PROBLEMS AND POLICY CONCERNS

Makers of public policy must grapple with a number of concerns that arise in the electronic environment. One concern has already been discussed in a previous section, namely the confusion over the legal models that apply to organizations engaging in electronic communication. In that domain, policymakers will have to sort out what will happen in the collision of legal regimes associated with different models and the boundaries (if any) at which public and private activity interact (as they might, for example, at the interface of private and public networks).

But other issues arise as well. For example, the question of responsibility for content goes beyond the different models (e.g., publisher, distributor) used by commercial information service providers for handling such responsibility. Questions in this area include the following: What content is acceptable and who decides what is acceptable? If penal statutes are violated, is criminal liability the bur-

---

Box 3.10
Ford v. Swarens, 1969

A computer error caused the Ford Motor Credit Company to believe that Mr. Swarens was delinquent in payments on his new car. On two occasions, the company's representatives visited Mr. Swarens and left after he showed canceled checks proving that he was not in fact delinquent. On the third visit, he threatened company representatives with a shotgun, after which his car was repossessed. In court, Mr. Swarens was awarded both compensatory and punitive damages, and the Court of Appeals said, "Ford explains that this whole incident occurred because of a mistake by a computer. Men feed data to a computer and men interpret the answer the computer spews forth. In this computerized age, the law must require that men in the use of computerized data regard those with whom they are dealing as more important than a perforation on a card. Trust in the infallibility of a computer is hardly a defense, when the opportunity to avoid the error is apparent and repeated."

SOURCE: *Ford Motor Credit Co. v. Swarens,* Court of Appeals of Kentucky, 447 S.W.2d 53, October 17, 1969.

---

den of the perpetrator or of the provider of the network service used to violate the statutes? What civil or tort liability applies to defamation or negligence? How do issues of pornography, indecency, and obscenity play out in the networked environment?

Anonymity and confidentiality pose other important concerns. On electronic networks, especially those associated with commercial information services, many people assume a different electronic "persona" that is not associated with a real person. Anonymity conveys both benefits and drawbacks. For example, history illustrates the use of anonymous communications to protect dissidents from the scrutiny or retaliation of autocratic governments. Even in recent years, the identification of a specific individual as a dissident (e.g., in the People's Republic of China) could mean imprisonment or even death; as a result, such societies often have forums (perhaps illegal but nonetheless public) in which anonymous protests against the existing regime can be posted. On a smaller scale, Murray Turoff, a professor of computer and information sciences and management at the New Jersey Institute of Technology, said his students (and executives) feel free to discuss on-line their problems at work only because they can do so anonymously.

The other side of anonymity was pointed out by Allan Adler, who argued that if users are granted true anonymity, "then there can be no justice system that will right the wrongs they commit, and there can be no 'rights' that people will be able to vindicate." In short, the justice system or law requires the identity of the person who caused the harm. Adler, along with Kiesler and Turoff, suggested that system operators keep users' identities confidential instead of anonymous and reserve the right to disclose the identity of a user who engages in injurious conduct. Today, the legal status of an anonymous message sent on an electronic network is not clear. Do network users have a right to refuse to disclose their identity? To prevent disclosure of their identity to a third party (i.e., one not associated with the network service provider)? To escape accountability and the law? Do users have to accept responsibility for their behavior?

A fourth concern is accessibility. Who can get into the system, and who can interconnect through which gateways? Policy questions include the following: Who can demand a right to interconnect or become a subscriber or participant? On what basis may service be denied to an end user or to an electronic connection? Is it necessary to mandate interconnection capability between various network services (perhaps in a manner analogous to the mandating of interconnection standards for telephone networks 50 years ago)?

A fifth area of concern is electronic commerce. Internet "netiquette" looks down on advertising messages that are publicly broadcast.[11] Yet it is easy to find Internet messages that look very much like

---

[11]For example, a commercial advertisement posted throughout Usenet, a network consisting primarily of bulletin boards devoted to various topics, has been widely condemned in the affected community. The Usenet community's response to unsolicited commercial advertising was largely to flood the advertiser's electronic mailbox with messages protesting the advertiser's action; other informal actions have been taken as well. See Peter Lewis, "Sneering at a Virtual Lynch Mob," *New York Times*, May 11, 1994, p. D-7.

More generally, this sentiment against commercial "advertising" seems to be rooted in at least two strains of thought. The first is the sentiment that advertising is an annoying pollutant, in the same generic class as junk mail. The second is that receiving e-mail consumes resources for which users must eventually pay and that it is inappropriate to make recipients of advertising incur those expenses if they have not affirmatively chosen to do so. The latter is in many ways analogous to many people's feelings about the unsought receipt of advertisements via fax, in which the recipient must pay for the paper consumed by the recipient's fax machine and the incoming telephone line is made unavailable for use by "legitimate" inbound fax transmissions. How these sentiments will evolve as the Internet is used more for commercial transactions remains to be seen.

classified advertising. Since the newspapers have spent a number of years in the courts and in Congress litigating and lobbying to prevent telecommunications firms from providing electronic advertising, these pseudo-advertising Internet messages constitute a very permissive interpretation of rules governing what is considered "appropriate use" of a publicly funded electronic space. As the Internet is opened to more commercial firms, battles over the extent and nature of "appropriate" commercial content can be expected. Questions that policymakers will have to address include: What are legitimate expectations of compensation for posters or authors of messages that may be valuable to others? What facts and circumstances define a copyright infringement on a network? How will potential users identify what information is proprietary? How will individuals monitor and enforce their rights and responsibilities with respect to billing, collection, and payment? What advertising messages will be permitted under what circumstances?

The definition of public domain in the area of commonality remains a problem. On a network, it is difficult to determine when something is in a private space and when something is in the public domain. Some observers warn that anything posted to a public mailing list on the Internet, or to a discussion forum or bulletin board on America OnLine or CompuServe, is a public statement that may be freely redistributed by other users. On the other hand, current copyright law does not require that the creator of a work explicitly insert a copyright notice in order to claim a copyright interest. So, upon viewing a statement on the Internet (for example) that does not contain a copyright notice, users are understandably uncertain about what rights they may be violating if they report the statement. It is arguably safer to assume that the writer is asserting a claim of copyright, and some "netiquette" rules direct the potential re-user to seek permission to report the message, but the legal murkiness remains.

Related is the question of what is subject to the Freedom of Information Act and the various sunshine laws. This is particularly important for advisory groups that may be using electronic mail or real-time conferences to transact their business. When can users expect to be operating in a confidential electronic environment? When must they invite the journalist or the public in? When does the electronic bit stream become a public record? When is a real-time electronic conference a "public meeting"?

The concern over "public" access and rights gives rise to an important definitional issue: the distinction between "public" and "private" enterprises or activities. In practice, there are at least four ways in which the "public versus private" distinction is used to distinguish networking activities: (1) the funding sources of the activity (public versus private funding); (2) the content carried by the networking activity (public information versus information not in the public domain); (3) whether the network activity is open (public) or closed (private); and (4) the ownership of the network activity (public versus private control).

Governance is particularly thorny. Society recognizes groups within it that are entitled to varying degrees of self-governance (e.g., condominium associations, corporations), although they are, of course, subject to a certain degree of external regulation as well. Today, electronic communities are making their own rules, applying them, and living by them. Under what circumstances can authorities of the government step in? (An important related question is what entity at what geographic level is best able to make and enforce what rules.) It is clear that government authorities can and do step in when actions taken within that community have harmful repercussions outside it or when members of the community request outside intervention, but the desirability of such intervention is subject to much argument. Questions over territorial jurisdiction are even more complex with global networks, because there are all sorts of complicated rules applicable to jurisdiction and extradition. Jurisdictional autonomy is a very rich area for lawyers and policymakers to sort out.[12]

Finally, what sanctions should apply when miscreant behavior occurs? Various states have been experimenting with different types of sanctions other than fines, jail, community service, and liability for damages. For example, as part of sentencing, judges have considerable latitude to impose restrictions on behavior, and so judges have proposed sentences (or conditions or parole or probation) that include the confiscation of the felon's computer equipment, prohibitions on the felon's use of computer equipment, and so on.

Informal sanctions have also been used outside the judicial system. Complaints directed to an offending user by others in the community are a particularly effective sanction against offenders, especially when the offender receives hundreds of complaints. By shunning offenders, members of the community apply a form of excommunication, an exercise of the traditional way communities have protected

---

[12]As one example, Ivan Orton, a prosecutor in Seattle, has taken 3 years to file suit against Richard Brandow, a Canadian who released what is called the peace virus into the "Mac world"—3 years just to get the case together. It is doubtful whether he will be able to win the case—and this case only involves Canada, a near neighbor with a not-too-dissimilar legal system. Imagine the complications when the parties involved are from nations with very dissimilar legal traditions!

their cultural and legal integrity. When a near-infinite number of venues can be established, it is possible to organize a group in a way that moves the entire group to another venue, leaving the offender without other communicants. How to determine the appropriate, acceptable, and effective sanctions to control miscreant behavior is an area that still needs careful analysis and development.

SUMMARY

The current legal system has evolved out of many decades of experience and precedent in dealing with various media for human expression and behavior. Electronic networks are a relatively new medium, but their novelty should not blind us to the simple fact that they do exist within the current legal context and thus that the current legal regime will have an effect on them. At the same time, their novelty does introduce complications and difficulties into this legal regime.

As electronic networks become more pervasive, a new legal regime will inevitably evolve, involving new law as well as new interpretations of current law. The first steps are being taken today in debates and arguments among users and providers—indeed, all participants in networked communities—as they try to sort out what values we share and what behavior they find acceptable. The ethical values that emerge from that process will be the foundation of a legal regime relevant to electronic networks that is designed to curb harmful behavior while preserving what is unique and distinctive about these new electronic communities.

4

Free Speech

The right of free speech enjoyed by Americans is rooted in the First Amendment, which states that "Congress shall make no law . . . abridging the freedom of speech . . . ." Nevertheless, the right to free speech is not entirely unfettered, and one's ability to speak whatever one likes can be legally limited under certain circumstances that depend on the nature of the speech and the communications medium in which that speech is expressed. The electronic environment, which gives every user access to a large audience and a virtually unlimited supply of information, poses particular challenges concerning free speech. This chapter summarizes a discussion of two free speech scenarios that were examined by a panel at CSTB's February 1993 forum.

NOTE: This chapter, and the three chapters following it, are based on the discussions held at the February 1993 forum described in the preface. As noted in the preface, the forum was intended to raise issues related to and associated with the rights and responsibilities of participants in networked communities as they arose in discussions of various hypothetical scenarios. Thus, Chapter 4 through 7 collectively have a more descriptive than analytical quality.

SCENARIO 1:
EXPLICIT PHOTOS ON A UNIVERSITY NETWORK

A large state university serves as a network "hub" for the state's high schools. The university itself is networked with every faculty member, staff member, and student having a network computer on his or her desk. The university also is connected to the Internet. A student electronically scans pictures of men and women in various sexual poses.

Issue: The Law as the Ultimate Authority

The university needs to consider how its policies are consistent with the law, because a state university exists in a jurisdiction that probably has indecency and obscenity laws, according to Allan Adler, a lawyer with Cohn and Marks. "We do not voluntarily submit ourselves to the law. It is the reality in which we live," he said.

Adler acknowledged, however, the common practical desire to reach consensus about behavior or conduct through negotiated policies and agreements. This is a typical approach in situations involving a communications medium, he said, where legal resolutions tend to be expensive and most parties get more than they bargained for in terms of restrictions on future conduct. Clearly, he said, there must be a set of social norms, whether defined by policy, contract, or law, and there must be some authority that enforces those norms; the ultimate authority is the law.

Assuming that the university in Scenario 1 communicated the network ground rules to users, their usage of the network would imply assent, Adler said. Assent is required in both contract and criminal law, by individuals in the former case and by society as a whole in the latter. That is, individuals must be notified of regulations that affect their conduct so that they have a fair opportunity to comply; if they do not assent to compliance, then they cannot fairly be held responsible for complying, he explained.

Responding to a suggestion that users could enforce rules themselves by employing screening devices, Adler argued that defamatory or fraudulent information would be difficult to informally identify and filter out, and the law would need to step in. For example, an individual about whom a defamatory story had been written might wish to prevent others from seeing the story, and thus would have to persuade an unknown universe of individuals to screen stories about him—clearly an imposing if not impossible task. "If the user is a participant in a system where either the user or some third party is defamed or has sustained damage to their reputation that affects them outside of the immediate electronic network on which they are operating, they are going to want some remedy for that," Adler said.[1]

As to whether the student, if disconnected, has any First Amendment rights, Larry Lessig, an assistant professor of constitutional law and contracts at the University of Chicago, noted that such rights would apply at a state university (though not necessarily at private universities). However, Lessig also contended that all universities habitually regulate speech. "If a student comes into my classroom and wants to talk about pornography when I want to talk about contracts, I tell the student, 'I'm going to fail you.' Now that is speech regulation," he asserted. Michael Godwin, staff counsel for the Electronic Frontier Foundation, concurred that the classroom is highly regulated and such regulation is appropriate based on the purpose of the classroom, but he raised the question of whether the university's electronic forum was more like a conversation on the block (in which freedom of speech guarantees do obtain) than like a lecture in a classroom.

Lessig warned that the legal community has few tools to make sense of behavior on electronic networks, in part because judges, lawyers, and legislators have little or no experience with that world. "I think the point is that we have very little understanding of how these principles that seem fundamental to us, like free speech, can apply in these various different worlds," he said.

Issue: The Need to Establish Rules and Educate Users

If the university is only now responding to the problems posed by Scenario 1, then it is already too late, according to Reid W. Crawford, legal advisor and interim vice president for external affairs at Iowa State University. He noted that networking issues should be considered within the university community before such problems arise, and any concerns should be shared with the connected high schools. He also said that discussions should begin very early in the process

---

[1]An alternative to legal intervention might be to require that any message about a person leave behind an electronic trail detailing where it was sent from, so that the person defamed can seek—and the remorseful author can send—a retraction or apology.

with university trustees, regents, legislators, the faculty senate, women's groups, civil liberties groups, and other concerned parties.

According to Crawford, this scenario would be an important public relations concern, because the university supports the network with public funds. Thus, he said, the matter should be handled preemptively through a quiet and nonpublic political and regulatory process managed by the university, involving consultation with the various constituencies in the university community. Crawford continued, "That is not the only step, but it's the first step that has to be taken so that you can deal with issues in a rational setting. Because if you think of it strictly as a legal issue, you can make all the arguments you want to about pornography, *Playboy*, *Playgirl*, or getting into the harder-core and illegal pornography. But if you cannot recognize the public relations issues, I don't think you will ever get to the substantive legal issues."

Since he made these remarks at the forum in 1993, Crawford's point has been underscored by scores of articles in the public press about "sex and the information superhighway."[2]

Lessig suggested that if users of a university network have an understanding that any subject is permissible, then it may be appropriate to have some technical means for segregating topics.[3] This idea was seconded by Murray Turoff, who noted that New Jersey Institute of Technology electronic forums supported the discussion of very controversial subjects in private conferences that are advertised in a directory. An affirmative choice must be made by the student before access to a private conference is granted.

Carrying forward the theme of individual responsibility for enforcement, David Hughes, a freedom-of-speech advocate and managing partner of Old Colorado City Communications, suggested that users who wished to screen out material they deem offensive could use a technological filter (akin to "Caller ID" for telephones), assum-

---

[2]See, for example, Amy Harmon, "The 'Seedy' Side of CD-ROMs," *Los Angeles Times*, November 29, 1993, p. A-1; John Schwartz, "Caution: Children at Play on Information Highway," *Washington Post*, November 28, 1993, pp. A-1 and A-26.

[3]This is an approach employed by America OnLine, which offers message boards and conferences for different topics. At the same time, the library community believes that any official scheme for "segregating" or labeling violates intellectual freedom. For example, the American Library Association's "Statements on Labeling" says that "labeling is the practice of describing or designating materials by affixing a prejudicial label and/or segregating them by a prejudicial system. The American Library Association opposes these means of predisposing people's attitudes toward library materials for the following reasons: Labeling is an attempt to prejudice attitudes and as such, it is a censor's tool. . . ."

ing such a device could be developed for electronic traffic.[4] This technological solution would be the equivalent of a porch—i.e., it would allow offensive messages to reach the door but not to enter the house. Still, other technical approaches for screening undesirable material may not be viable in the electronic environment. For example, broadcast media have often resorted to segregating material not intended for children by broadcasting such material late at night; in an electronic environment in which the material is available 24 hours per day, such an approach becomes more difficult to implement.

The first concern of George Perry, vice president and general counsel for the Prodigy Services Company, with regard to Scenario 1 was whether any laws were broken; he was not sure, noting that the answer might depend on the nature of the pictures. Therefore, the real question becomes the nature of the relationship among the user, the provider, and, perhaps, the victim, he said. Perry did not see a freedom-of-speech issue in the scenario, arguing that speech has never been entirely free.

Given that electronic networks are a new medium with very few commonly accepted rules of behavior, Perry emphasized the need for providers to establish them. "I think it is absolutely critical that operators of these systems, whether they are universities or commercial operations, establish their rules . . . . I don't think that every network in which you express yourself has to have the same rules, but you've got to have some rules; otherwise the place is just going to crumble."

A view that went farther was espoused by Hughes. He contended that the primary responsibility in handling the explicit photographs lay with the individual and that the university had no responsibility. If the university responded to the incident by disconnecting the student from the network, then the student should respond by demanding his or her freedom of electronic speech, Hughes asserted.

---

[4]The technological feasibility of an electronic filter is not at all a given, although some prototypes are being developed. It is easy to imagine a filter that would delete or suppress text-based messages that contained, for example, certain four-letter words; it would be far more difficult, however, to develop a filter that could screen out bit-mapped graphic images of humans in sexual poses. (One could, of course, screen authors' names and subject keywords.) Moreover, any such filter would require a user to specify with some precision what he or she found offensive, a feat not all individuals could accomplish. Example-based learning techniques may enable a program to derive a filter based on a few samples of undesirable material, but whether such techniques are feasible and applicable on a large operational scale is the subject of considerable technical argument.

He based this argument on the premise that the digitized photographs constitute speech rather than property, and that freedom of speech is the overriding principle. Methods of handling problems on electronic networks should evolve from that foundation, Hughes argued, rather than from university policies or social norms developed for other communications media.

In Hughes's view, the university's only responsibility in Scenario 1 is to educate students and the public. He asserted that members of the public who lack experience with electronic networks—including parents and newspaper reporters—are unable to make sound, objective judgments about such situations. On his own network, Hughes educates and influences his user population to adhere to an ethic. "We discuss that ethic, and out comes socially responsible behavior—not by imposing external authority on them. I never delete, on my system, a piece of mail—never, no matter how obtrusive. I handle it technologically sometimes [by masking the appearance of the message] so that it is free speech, and we have a discussion to the point of responsible group behavior." Others argue that masking the appearance of the message is itself a form of censorship.

In closing discussion of Scenario 1, moderator Henry Perritt noted that panel members generally agreed on the importance of establishing rules under which electronic forums operate. Rules need not be the same from forum to forum, as Perry pointed out, but they need to be explicit and give important consideration to the views of stakeholders—operators and users. Perritt pointed out that the disagreements were over the extent (if any) to which other mechanisms were needed to enforce those rules.

SCENARIO 2:
NEGATIVE COMMENTS HARM A THIRD PARTY

In an investment forum bulletin board hosted by a commercial network service provider, several users are discussing the merits of investing in XYZ Corp., a "penny stock" whose price can fluctuate widely on relatively small trading volume. John, a regular user of the bulletin board, has gained some credibility with other users for his stock picks. He posts a note on the bulletin board that says: "I was heavy in this stock 4 months ago, but sold most of my holdings last month. The company is out of cash, and sales are in the tank. Inside management is waiting for the stock to go up a quarter point to dump some big positions." The next day XYZ's shares fall precipitously on heavy trading.

Issue: Provider Responsibility and Liability

Lessig argued that because the First Amendment applies less forcefully in the context of commercial speech, it may be better to first ask to what extent it is reasonable to make individuals (in this case, the provider) seek out and screen information to avoid harming others. Lessig outlined two points relevant to the problem.

First, he offered the analogy of electronic networks functioning as the *National Enquirer* of cyberspace, in that no one could reasonably rely on anything that was said. Thus, "the person has no claim they had been harmed, because they shouldn't have relied upon [the information], and nobody should have relied upon it." That is not to say, however, that speakers should be immune from liability, Lessig added. It is only when individuals feel responsible for their words and actions that others begin to give them credibility, he noted. Thus, if the network is to gain credibility there must be some responsibility.

Adler was disturbed by the *National Enquirer* analogy, arguing that credibility is essential if electronic networks are to evolve into the marketplace of the future. Trying to ward off liability by warning users to distrust the network is the wrong approach, he said; users must feel comfortable knowing that they do not routinely risk any type of injury and that injuries, when sustained, can be redressed. Moreover, the First Amendment has been linked to Oliver Wendell Holmes's notion that the marketplace is where truth will prevail through a free competition of ideas, Adler added.[5]

Lessig's second point was that what is reasonable liability depends primarily on technology. If there were a simple way to search the bulletin board for harmful information, then common law courts

---

[5]Specifically, Justice Holmes wrote that "when men have realized that time has upset many fighting faiths, they may come to believe even more than they believe the very foundations of their own conduct that the ultimate good desired is better reached by free trade in ideas—that the best test of truth is the power of the thought to get itself accepted in the competition of the market, and that truth is the only ground upon which their wishes safely can be carried out. That at any rate is the theory of our Constitution. It is an experiment, as all life is an experiment." *See Abrams v. United States*, 250 U.S. 616, 630 (1919).

might find it reasonable to impose that duty on providers, he said.[6] However, emphasizing his earlier point, Lessig pointed out that common law is made by judges, who may be unfamiliar with electronic networks. "The biggest and, I think, most frightening thing about regulation in this area is that the people doing the regulation have no experience at all," he said. "They have no conception of what cyberspace looks like or even feels like."

There was some dispute on this point. Marc Rotenberg, a privacy advocate with Computer Professionals for Social Responsibility, suggested that *Cubby Inc. v. CompuServe Inc.* (described in Chapter 3) was a landmark free speech case that demonstrated that judges are beginning to understand electronic networks. Rotenberg sympathized with CompuServe's argument that it acted as a distributor, not a publisher, and that it did not know and had no reason to know of the statements in question. The court agreed, emphasizing the First Amendment and saying that CompuServe deserved the same protection as a traditional news vendor. Rotenberg was most enthusiastic about the *Cubby* decision, calling it "wonderful" and "sensible." He said the court appeared to be promoting electronic networks as an information resource by limiting but not eliminating liability for providers. Moderator Perritt suggested that even if judges didn't understand the new technology, they could be educated about specialized subjects through the testimony of expert witnesses and amicus briefs and that such approaches could be encouraged for cases involving electronic networks.

Others, however, warned that the decision establishes troubling precedents. Adler was disturbed by the finding that CompuServe was not responsible because it had little or no editorial control. In fact, the provider did have the technological ability to exercise control but chose not to do so, placing that responsibility on a contractor, Adler noted; he further wondered whether publishers should be allowed to pass liability on to a contractor simply by declaring themselves to be distributors and thus lowering their liability. "So when

---

[6]If the cost of searching the bulletin board for harmful postings were less than the value of the damage caused by information likely to be on the bulletin board, common law would hold the bulletin board owner liable for not undertaking a search if damaging information were indeed found on the board. (More formally, the cost comparison would be between the cost of the search and the expected value of the damage associated with harmful information (i.e., the product of the likelihood that the information will be found on the board times the damage it would cause if it were indeed found).) See Richard Posner, *Economic Analysis of Law*, 4th Ed., Little, Brown, Boston, Mass., 1992, pp. 163-169.

we are talking about rights and responsibilities in this world, one of the things we have to consider is, do the responsibilities flow from whether or not you actually have the capability, whether or not it is feasible for you to exercise control, or whether or not you choose to place yourself in that position of exercising control?" In addition, rights and responsibilities flow from a social decision regarding whether it is beneficial to grant individuals the rights or to saddle them with the responsibilities.

The central problem with the *Cubby* decision, Adler said, is lack of clarity, or failure to distinguish among the various electronic services and formats. Adler said he feared that some readers of the decision would conclude that solicitation of criminal activity, defamation, and other crimes or torts could be carried out on electronic networks without liability.

Hughes said he was frightened by the notion that a provider's capability to review materials translates into an obligation to do so. It is the provider that defines its services and thus its obligations, he argued. (Of course, a provider never has complete freedom, as it is subject to the laws of the jurisdiction in which it operates.) Hughes said further that even a single system may have multiple roles. For example, the Prodigy Services Company is a publisher, which implies some review of content; but the service also carries free speech, for which it should not be held accountable, according to Hughes.

The point regarding multiple roles was reinforced by Davis Foulger from IBM, who argued that different types of computer-mediated communications (e.g., electronic newsletters, conferences, and mail) may carry different types of responsibility. Electronic newsletters, he suggested, may be entirely analogous to print-based newsletters, with all of the liabilities of the latter carrying over to the former, whereas an unmoderated conferencing forum may carry fewer responsibilities. Given the flexibility offered by electronic networks to define the type of communication, Hughes and Foulger agreed that the self-definition ought to be the element that defines liability; for example, in the case of a commercial network whose contractual agreement with users declares that the network owns all data on its system, the network should be subject to all of the legal mechanisms used to hold an individual liable for that data.

Hughes went on to note a further complication to the self-definition process: the extensive interconnection between networks means that a given network may be unable to control the input to it. Hughes asked: "To what extent am I accountable for what someone else says on another system that happens to be displayed on mine?"

Godwin argued that *Cubby* posed a "knew" or "should have known" standard of liability for defamation, and he thought that the decision was proper. But Godwin also argued that the case does not imply that liability results from a complete failure to exercise discretion or a decision not to exercise discretion. He noted that "*Cubby* is based on *California v. Smith*,[7] [in which] bookstores were held to be not responsible for monitoring and not responsible for the specific contents of their books [even though] bookstores can exercise discretion about what they carry." He further argued that a broad kind of discretion need not necessarily result in liability; such a freedom from liability would be important to forum operators, who need to be able to shape the character of their forums but who also want to avoid liability for the specific contents of those forums.

Providers were satisfied with the *Cubby* decision but said it leaves some questions unanswered. Perry, while pleased that CompuServe was not held liable, said a provider's obligation remains unclear. For instance, if a system is large, with perhaps 50 million messages posted each year, how far should the operator have to go in investigating allegations about a piece of information? "Do I have an obligation to go find that thing in the 50 million [notes]? . . . Once I find it, do I have an obligation to go out and discover the facts as to whether or not what the individual was saying was true . . . ?"

Perry argued that, in the scenario he just outlined, both a publisher and a distributor have an obligation to determine the facts. This responsibility is clearer in the case of a publisher, he noted; in the *Cubby* case, the court found that if a distributor is notified of a problem, then it, too, faces some liability unless it takes action. Moreover, he warned that "it is very dangerous to make the publisher/distributor distinction when it comes particularly to commercial operators, because in fact they are a wide spectrum of different kinds of beasts. At one end they may very well be publishers. On another end they may be pure distributors. On another end they may be none of either. So I think it's dangerous for us to take those preexisting analogies and try to apply the law that we have today that applies to those areas, to this new medium."

In general, liability protection models from other media may not be appropriate on electronic networks, Adler said, noting that the

---

[7]In this decision (361 U.S. 147 (1959)) the U.S. Supreme Court held that a distributor's lack of editorial control precluded states from holding the distributor strictly liable for publication contents (A.J. Sassan, "*Cubby Inc. v. CompuServe Inc.*—Comparing Apples to Oranges: The Need for a New Media Classification," *Software Law Journal*, Vol. V, 1992).

distributor model suggests that "the best way to avoid any possible liability is to exercise the least control. [But] I'm not sure that is socially responsible." Rule by law is not necessarily so bad, he said, pointing out that, historically, citizens have not objected to laws per se, but rather to the arbitrary exercise of law—law without participation and consent.

As to how liability should be applied, Adler emphasized the distinction between a common carrier and a service provider: the former is virtually immune from liability because it is legally required to provide equal access to all users without editorial control over content, whereas the latter can be held responsible if it is notified of a problem and does nothing to eliminate the continuing harm. It is not clear which definition applies in Scenario 2.

Issue: User Responsibility and Liability

Liability for defamation is a critical issue in the electronic environment. In common law, the question of defamation rested on the truth or falsity of a statement about an individual. However, in light of First Amendment considerations, the Supreme Court has focused on the degree of fault that can be attributed to the speaker, Adler said. Of course, if the speaker (or poster) of a defamatory message is truly anonymous (i.e., if it is genuinely impossible to determine the identity of the speaker with certainty, as might be the case if the message originated on another network, for example), then the matter ends there, and no party can be held liable.

True anonymity currently is rare on electronic networks.[8] In many more cases, the true identities of speakers are confidential (i.e., the identities of speakers are withheld as a matter of policy on the part of the service provider, although the provider does in fact know these true identities). In such cases, Adler maintained, "the question is whether the service provider is willing to accept the liability for the harm that is caused by maintaining the promise of confidentiality, or whether . . . there is a balance . . . that says there are compelling interests which outweigh the values of the promise of confidentiality and require disclosure."

But Perry and Hughes took a somewhat different tack. Perry argued that a user's liability on a bulletin board ought to be the same

---

[8]Anonymous use of electronic networks is nevertheless expected to increase. Even today, there are so-called "anonymous remailers" that accept e-mail messages and forward them to their intended recipient stripped of any identifying information.

as in any other circumstance. Hughes agreed, saying that John's note in Scenario 2 was an individual act of irresponsible speech, for which the provider was not responsible. Both suggested that under the circumstances of Scenario 2, the provider should not have to bear the liability to which Adler referred. Hughes further argued that when identities are kept confidential as a policy choice on the part of the provider, a complainant should seek the assistance of law enforcement authorities and show probable cause for issuance of a search warrant (in a criminal case) or a subpoena (in a civil case) that would compel the provider to provide the sender's identity. In this case, the decision concerning whether to divulge the speaker's identity does not rest with the provider but with law enforcement authorities.

If the provider in Scenario 2 maintained the confidentiality of the originator of the communication, thus shielding the only potential defendant, then the provider still would not be obligated to report a violation to law enforcement authorities, Hughes said, citing the Electronic Communications Privacy Act of 1986 (Public Law 99-508). However, he said, the provider has some fundamental ethical responsibility. "Some things are right and some things are wrong," Hughes argued.

DISCUSSION AND COMMON THEMES

The balance between free speech and other values is tested regularly, and the regulation of speech comes in many forms. Even in academia, usually regarded as the bastion of free speech, users sometimes are banned from networks. Carl Kadie, a graduate student in computer science at the University of Illinois and moderator of the academic-freedom mailing list on the Internet, cited the example of an Iowa State University student being expelled from the campus network for copying materials from an erotic forum into an open forum meant for discussion of newsgroups and newsgroup policy. The expulsion was lifted after protests from Internet and campus network users, but access to the erotic forum remains restricted. Kadie went on to argue that because many universities are state universities and because many parts of the Internet are owned or leased by federal or state government, lawsuits could be filed in these cases based on the First Amendment,[9] although the law is weaker for cases involving library policy on selection (or, more to the point, exclu-

---

[9]The First Amendment applies only to public forums, which can result from government funding or a dedication to public use (e.g., an airport or a shopping mall even if privately funded), and different types of speech may have different levels of protection.

sion) of electronic resources. "I think many of these arguments have
to be won or lost on the moral argument and [by] appealing to freedom of expression—academic freedom—and can't depend so much
on legal protections," Kadie said.

Still, users generally can be more outspoken on university networks than anywhere else. Kadie expressed the hope that those who
determine rules for behavior on electronic networks will "learn from
the experience (and, hopefully, wisdom) codified in long-standing
academic policies and principles . . . . I don't think academia should
be the only forum for free speech. . . . I would hope that technical
solutions such as 'kill files'[10] and the ability to create new forums [as
is done in Santa Monica] would be enough to have people [on non-
academic networks] regulate themselves."

It may be difficult to transfer this principle of openness to other
arenas. For instance, corporate executives may want to control postings
of material relevant to company business. Economic considerations
also may argue for the regulation of speech at times. This is particularly true for commercial information providers, as clients demand to
be insulated from certain types of content, noted Perry. Perry said,
"There are different environments in which you have to deal with
the same set of problems, maybe with a different view and a different historical and traditional background." These differences in perspective can lead to conflicting views about what speech is or is not
appropriate for public viewing or exposure.

Finally, it was argued that free speech was not possible when one
was denied access to the electronic environment. As Sara Kiesler
noted, "The purest form of censorship is absence of access. If you
can't have access to a network at all, then you are completely censored from that forum. Therefore, we have to know who has access,
and, even for those who have physical access, who's being driven
off."

There was broad agreement among panel participants that free
speech is not an absolute right to be exercised under all circumstances.
The relevant issue for free speech is what circumstances justify the
uses of which mechanisms to discourage and/or suppress certain
types of speech. The nature of the circumstances in which free speech
should be discouraged is a matter of political and social debate, but it
is clear that policymakers have a variety of such mechanisms at their
disposal to discourage or suppress expression:

---

[10]A "kill file" is a list of users (i.e., network accounts) whose messages are to be
deleted automatically from the set of messages shown to the owner of the kill file.

- *Educate and persuade:*   At a minimum, most people agree that an important step involves persuasion and education, relying on voluntary means to dissuade people from saying things that are arguably harmful, objectionable, or offensive to others.
- *Rely on contractual provisions:* Someone agrees, as a condition of use, to abide by regulations regarding the content of speech.
- *Weigh political considerations:* An institution may wish to weigh how it will be seen in the eyes of its relevant public in determining the nature of its response.
- *Rely on market mechanisms:* Grass-roots pressure on suppliers of information services often forces change because suppliers fear losing the business of those complaining.
- *Explicitly rely on First Amendment freedoms:* A state university may have far fewer options for regulating speech due to the fact that it can be regarded as an arm of government.

How are these issues different in the electronic networked environment? Lessig asserted that

> in ordinary life, social norms are created in a context where other things besides speech are going on, things such as exclusion, anger, and the impact of local geographies. . . . [These are the] sorts of things that help the process of norm creation in a speech context. [But] what makes electronic networks so difficult from the perspective of creating and molding norms is that the interactive human behavior on these networks is mostly if not entirely pure speech. From the constitutional perspective, this is the first environment in which society has had to face the problem of creating and changing norms when the only thing it is doing is trying to regulate speech.

The steering committee generally concurred with this assessment, concluding that

- Networks offer a greater degree of anonymity than is possible for speakers under other circumstances.
- Networks enable communications to very large audiences at relatively low cost as compared to traditional media.
- Networks are a relatively new medium for communications, and there are few precedents and little experience to guide the behavior of individuals using this medium. As a result,
    — Speakers are less familiar with a sense of appropriateness and ethics here (treating a big megaphone as though it were a smaller one); and
    — Policymakers are less confident in this domain.

5

Electronic Vandalism

The extent of electronic vandalism and other computer-related crime is unknown. Annual losses in the United States due to computer crime have been estimated at $2 million to $730 million, but these figures are suspect; several studies have estimated that only 1 percent of all computer crime is detected, and the Federal Bureau of Investigation (FBI) estimates that of crimes that are detected, only 14 percent are reported.[1] In any case, it is clear that security on electronic networks is a growing concern.[2] Antivirus programs have become almost ubiquitous, and, as Eberhard Wunderlich of AT&T noted at the November workshop, there is a trend toward moving beyond user passwords to smart cards for access control, encryption, and electronic signatures. In the first half of 1994, concerns for security on the Internet were prompted by a rash of penetration attempts, many of which were successful.[3]

But technology cannot guarantee absolute security. Forum par-

---

[1]C.D. Chen, "Computer Crime and the Computer Fraud and Abuse Act of 1986," *Computer Law Journal*, Vol. X(1), pp. 71-86, Winter 1990.

[2]A classic case of network penetration is that of the Hanover Hacker, recounted by Clifford Stoll in *The Cuckoo's Egg: Tracking a Spy Through the Maze of Computer Espionage*, Doubleday, New York, 1989.

[3]Peter H. Lewis, "Hackers on Internet Posing Security Risks, Experts Say," *New York Times*, July 21, 1994, p. A-1.

ticipants indicated, for instance, that a system operator may have difficulty detecting a fraudulent (i.e., stolen) user identification, particularly one forged by an expert. As Mitchell D. Kapor, co-founder and chair of the Electronic Frontier Foundation, said, "There's always some tiny fraction of the elite that is, in fact, technically capable of inflicting huge amounts of damage . . . ."

For those cases in which damage does occur, some remedy may be found in the law, though even this remedy is subject to the caveat that its efficacy depends on the reality of enforcement. If a district attorney refuses to prosecute or the civil courts take too long to handle a lawsuit, the determination of trespass or theft becomes an academic discussion.

SCENARIO 1: VIRUS DAMAGES BULLETIN BOARD

A computer club at a local high school sets up a dial-in bulletin board, using equipment bought for the club by a banker whose son is club president. The bulletin board is set up at the club president's home, and it can exchange messages with other bulletin boards across North America. The banker also has a computer system for working at home that is tied directly into the club's computer; the banker's computer is used to write a public newsletter for his bank. The telephone number of the bulletin board is distributed through a national magazine, and over time, the following activities are taking place on the bulletin board, although no club members are involved in any of these activities:

- Stolen credit card numbers are posted;
- Hate messages are sent to Canada, where such messages are illegal;
- A program is posted in a public space by Joe, a nonmember, and others download the program and discover that it contains a virus that causes considerable damage; and
- A second program is posted that is designed to disrupt network services when run on a computer connected to a national network such as the Internet or CompuServe.

Issue: Criminal Liability

Club members have no criminal liability for failing to monitor the activities in this scenario, said Scott Charney, chief of the Computer Crime Unit at the Justice Department. If a potential criminal case involving a computer warranted an investigation, both the law and sensible policy would dictate use of the least intrusive means, such as a subpoena, to obtain information, he said. However, the answers to several questions might alter this response. The first question is how prosecutors learned of the problem; the club members themselves might have reported it, for example, thereby inviting an investigation.

The second and more complicated question involves intent. "Suppose, for example, that the people running this board knew about the illegal access codes and were actually fostering it and encouraging it," Charney said. "That doesn't mean the [bulletin board] is any less valid, but those running it may not be innocent third parties anymore. And how do you investigate that and get the evidence you need without infringing on the newsletter, which is protected under the First Amendment to the Constitution?"

If they pursued the virus case in Scenario 1, federal prosecutors would not seize equipment or search the bank computers, because neither the bulletin board nor the bank is involved in criminal activity, Charney said. In fact, it is not clear whether there is any criminal case at all; if the bulletin board is open to the public, then Joe, even as a nonmember, has access authority and hence no criminal liability, because the statute under which criminal liability is imposed requires unauthorized access.

Issue: Deficiencies in the Laws

Mark Rasch, an attorney with the firm Arent, Fox, Kintner, Plotkin, and Kahn, said Joe's inappropriate behavior has two elements: possible unauthorized access to the bulletin board and uploading the computer virus. Since Joe is arguably an "authorized" user (anyone could be an "authorized" user of a public bulletin board system, and 18 U.S.C. Section 1030 simply prohibits unauthorized access), Joe might well escape any liability at all. As for the uploading of the computer virus, "Prosecutors must prove Joe's intent. Why did he do it? The way the federal law is structured right now, if Joe accesses with authorization, even with intent to destroy and damage all the information in that computer, the computer crime statute says he is not guilty of a crime, because it is his initial access [that the law focuses on]."[4]

---

[4]Note that Joe may still be liable to prosecution under the applicable state's criminal law, if any.

Rasch suggested that legislators based the law on an imperfect analogy—an office break-in, where no matter what intruders do in the office, they are guilty at minimum of trespassing. Although the analogy falls short, Rasch said lawmakers have relied on such analogies because of the lack of common experience in the environment of electronic networks. "When we see crimes in society, we can relate to them as criminal acts because we have a common experience about what is a crime and what is not a crime," he said. "And when we deal in cyberspace and we deal in the electronic environment, we don't have those experiences. We don't have those common human judgments. We don't have that ethical standard to fall back on. So then we end up . . . trying to draw analogies."

Kent Alexander, now U.S. attorney for the Northern District of Georgia, said Joe has extensive criminal and civil liability but that prosecuting him would be a challenge. Alexander said the Computer Fraud and Abuse Act of 1986 seldom applies in real cases, adding that prosecutors often depend on broader laws such as the wire fraud statute (18 U.S.C. Section 1343), which applies if a telephone is used in a scheme to defraud. Alexander and Rasch both said a federal destruction-of-property statute is needed.

That prospect prompted some caution. For example, Marc Rotenberg and Michael Godwin argued that although it was appropriate to criminalize undesirable behavior such as destruction of property and computer trespass, it was unnecessary to specify what the instrumentality of such behavior was. Godwin said the law generally approaches computers as if they were "inherently dangerous" in ways that other general-purpose tools, such as hammers, are not; he argued that this legal approach infringes on freedom of speech and that computer use should not be overregulated in this manner. Still, Alexander insisted that some computer-related laws are needed, because "cyberspace is . . . what we are going to live in." Rasch said, "The reason we need specific computer crime statutes is because analogies don't actually work perfectly," although he cautioned that it would be preferable to target a potential miscreant's overall strategy rather than a precise mechanism, because a vandal might be able to "engineer around" a defined mechanism.

Litigation specialist Thomas Guidoboni, of the firm Bonner and O'Connell, agreed with other speakers that the Computer Fraud and Abuse Act is flawed. He argued that victims should be able to recover damages through the civil system but that criminalizing and jailing hackers does not benefit victims, other than by serving as a deterrent. (The statute specifies punishment consisting of a fine and/ or imprisonment.) He contended that, although the Congress prohibited computer trespassing, it did not address worse wrongs (such as the malicious destruction of computer files): "They need to be educated . . . . [T]hey don't really know what to prohibit."

Charney said the federal law should specify the type(s) of prohibited network intrusion. He noted that the Justice Department has asked the Congress to amend the Computer Fraud and Abuse Act because the ethics of computer usage have evolved and users' roles have changed since 1986. "It's better to say, 'Look, if you upload a virus, even if you are authorized [to access the computer] . . . intending to damage, without authority, other people's material, that's going to be criminal.' That way, everybody has plenty of notice and you've got a statute that directly applies to the conduct." Such a statute should not criminalize legitimate work with viruses, such as in the research community, he added.

Godwin noted that federal law doesn't require that *intent to damage* be established, and he suggested it is bad policy for criminal laws to ignore intent. He said criminal statutes generally, and rightly, focus on the intent or mental state of the defendant, because "we try to criminalize people for setting out to do bad things, not for unwittingly doing bad things." Joe's potential criminal liability for putting out a virus that disrupts network service or that damages a "federal interest" computer[5] may not reflect the proper balance between civil and criminal sanctions, Godwin suggested, because the law now "makes no distinction between someone who accidentally causes damage and perhaps a terrorist who deliberately causes damage to vital systems." Rasch countered by arguing that even if the statute itself makes no such distinction, it is possible to make distinctions in other ways, such as in sentencing; he cited the sentence in the *Morris* case as an example (Box 5.1), arguing that Morris was not imprisoned because the damage he caused was accidental.

Godwin noted that Morris had authorized access to the computer from which he launched the virus. Guidoboni, who defended Morris, had argued unsuccessfully that his client, as a member of the Internet community, had authorized access to all computers on the system. Guidoboni pointed out at the forum that "authorization" and "access" can be difficult to define for the Internet. Guidoboni also warned against attempting to curb computer vandalism through "foolish"

---

[5]As defined in the Computer Fraud and Abuse Act of 1986, a federal interest computer is a computer used exclusively or part-time by the U.S. government or a financial institution, where the offense affects the government's operation of the computer. The term also applies if two or more computers, not all of which are located in the same state, are used in committing the offense.

Box 5.1
***United States v. Robert T. Morris,* 1991**

Robert Morris was one of the first to be prosecuted under the Computer Fraud and Abuse Act of 1986. As a Cornell University graduate student, Morris wrote and released onto the Internet a "worm" program designed to propagate itself automatically without harm to the host computers, but due to a flaw in its design, the Morris worm caused thousands of computers nationwide to crash on November 2-3, 1988. By itself, the worm did no permanent damage, although the estimated cost of rectifying its effects at each location involved ranged from $200 to $53,000. As a graduate student in the Cornell University computer science department, he had full and legal access to all the department's computing facilities; these facilities were connected to the Internet.

Although Morris' defense claimed the worm was a benign test of computer security that went wrong, Morris was convicted on the grounds that he intentionally accessed remote computers without authorization. He was sentenced to 3 years probation, fined $10,050, and ordered to perform 400 hours of community service, a sentence far less than the maximum permissible under law. The conviction later was upheld by a federal appeals court, and review was denied by the Supreme Court.

SOURCE: *United States v. Robert T. Morris,* No. 89-CR-139 (N.D.N.Y.), aff'd 928 F.2d 504 (2nd Cir. 1991), cert. denied, 112 S. Ct. 72 (1991).

---

censorship, noting that the government refused to release the "cure" for the Morris virus, for fear of hackers obtaining it. "So of course a lot of you people on the Internet couldn't fix what was wrong, because our government didn't want to give you the code to fix it . . . . [It was their effort to] prevent a perceived wrong that probably did more harm in the end."

Issue: Civil Obligations and Liabilities

Civil liabilities arise from the general principle in common law that a person who has been wronged by another person is entitled to compensation for that wrong from that other person. Charney suggested that Joe might have considerable civil liability if the damage he caused was extensive. Alexander said a civil case against Joe probably would not hold up, assuming there were no federal agents or state police involved, because hackers usually are judgment-proof.

(Hackers are usually judgment-proof because they are typically young people who have no assets to seize.) Conversely, although the father in Scenario 1 has not committed a criminal act, he probably will be sued because he and his bank are the only parties with any money, Guidoboni observed. Whether there is a valid case is doubtful, he added. Godwin noted that civil liability depends little, if at all, on the intentions of a defendant.

Godwin argued that "to a large extent, many of the issues regarding liability are not criminal law issues at all. We have a large body of civil law devoted to negligence obligations, obligations to be nonnegligent in maintaining other kinds of spaces, be it your yard or your business." He contended that, ultimately, individuals who maintain forums will have some duties of care that arise from civil obligations. He cautioned, however, that because an on-line bulletin board is a forum for speech, whatever civil obligations are imposed on a system operator must be consistent with the First Amendment.[6] He further contended that distribution of virus codes or information, and general discussion of viruses, are protected by the First Amendment. "It's always very troubling to discover that someone's teenaged son or daughter has access to a potentially damaging bit of computer code," he said. "But one of the choices that we have made in living in a free society is to say we're going to allow people to be a source of a lot of risk. We're going to risk having people who have dangerous information."

SCENARIO 2: MULTISYSTEM INTRUDER DAMAGES FILES

Computer A is penetrated by an unauthorized user. The intruder uses Computer A to reach Computer B, where he causes extensive damage to files. The operator of Computer A monitors the keystrokes of the intruder, who eventually is caught. The intruder claims that the operator violated the wiretapping statute.[7] Computer A did

---

[6]As private parties, Prodigy, America OnLine, CompuServe, or GEnie are not subject to the First Amendment in the sense that they have the right to establish the rules that apply to the use of their services as a contractual matter and to refuse to provide service to those subscribers who do not adhere to such rules. On the other hand, actions that can be legally taken by a governmental entity must be consistent with the First Amendment; to the extent that such actions are required to enforce a contract, all parties to the contract are affected by the First Amendment.

[7] The wiretap statute is *United States Code Annotated*, Title 18, Sections 2510-2521. To

not display a warning message regarding unauthorized access because its operators forgot to install such a message, while Computer B did not display a warning message on the premise that anyone accessing it from Computer A would be an authorized user.

Issue: Trespassing and Theft

Scenario 2 is principally a trespassing case, and it illustrates the need for computer crime statutes, Rasch said. Computer trespassing is abhorrent, he asserted, even if no damage is done, because it destroys trust in the system. The crime is different in character from a home or office break-in because the distrust lingers. "When you are a user, and you see somebody without authorization on your system, you don't know that he didn't do anything, and it can take you a very long time to come to the conclusion that he didn't do anything," Rasch said. "It would be like having a calculator that 1 out of every 100,000 times adds the numbers up incorrectly."

Rasch argued that although trespassing is a crime in itself, monetary damage also should be calculated, to adjust the level of the offense. At the same time, he raised the issue of whether stealing electronic information should be criminalized at all. "If I steal the files in your office, you know I've stolen them. If, on the other hand, I simply copy them, the law is not clear on whether I've stolen something. In a computer environment, I can steal all the information in your computer and you won't know it, and you'll still have all the information."[8]

Although the federal statute does not criminalize reckless or negligent actions, Rasch said negligence will become an increasingly common

---

use wiretaps for collecting information about the content of communications, law enforcement officials must obtain a court order by demonstrating probable cause and necessity. However, a party to a communication can consent to its interception. Federal policy holds that system operators do not have a privileged status allowing them to monitor communications, and so federal agencies place notices on computer systems warning hackers that they are subject to monitoring, according to Patrick Lanthier, director of public policy and technology for Pacific Bell. Different policies may obtain under state law.

[8]There is a long legal tradition of protecting intellectual property, which is itself a legal construction that vests in certain people property interests in the intangible (for example, the law grants authors copyright protection and companies trade secret protection). However, the boundaries that separate criminal versus civil liability for violating these laws are fuzzier than those for governing theft of tangible property.

basis for civil litigation. As to whether reckless actions should be criminalized, he acknowledged that this approach would help prosecute wrongdoing and encourage users to be careful, but he noted that the problem of proving a defendant's mental state would persist. He dismissed claims that intent can be difficult to prove, saying, "A person is presumed to intend the natural and probable consequences of his actions, so generally you can prove intent."

Godwin argued that the phrase "stealing information" implies that information is property, when in fact an argument can be made that information should be assumed not to be property.[9] To become property and be subject to theft, information must meet certain tests, such as those imposed by copyright or trade secret laws, he said. He agreed with prosecutor Charney that privacy is the key issue in computer intrusion, just as it is in wiretapping. "I think it's appropriate sometimes to criminalize intrusion and the examination and even the copying of data as perhaps the violation of a privacy interest," Godwin said. "But treating it like a theft crime, I think, doesn't help the public consciousness of . . . the true nature of these crimes."

While he supported the use of incentives to curb user negligence, Godwin questioned whether criminal law is the right tool. Users already have incentives to secure their systems and refrain from distributing damaging software; whether these incentives are adequate is the question, he said.

Godwin argued that criminal behavior on electronic networks should not be defined until "we actually have a sense of what the social norms are. . . . Many of us just say it's obvious that intruding in a computer is like intruding in your home," he said. "Believe me, if it were that obvious, then many of the people who have done computer intrusions would never have done it, because these people have never walked [without permission] into other people's homes."

Issue: Determining Damage

Assuming the two computers are in different states (i.e., a federal interest computer was involved pursuant to 18 U.S.C. Section 1030(e)(2)),

---

[9]The question of whether or not information should be or can be regarded as property is subject to considerable debate. What is undeniable is that certain types of information and perhaps certain pieces of information (i.e., specific instantiations of those types) are generally regarded by society at large as being more appropriately shared or appropriately private than other types (and perhaps pieces) of information. Put differently, agreement on the principles that differentiate these categories could not be expected to result if a group of randomly chosen "reasonable people" were asked to articulate such principles.

and the intruder is not authorized to access either Computer A or Computer B, the intruder could be prosecuted under the Computer Fraud and Abuse Act (Section 1030(a)(5)) if damage totaled $1,000 or more,[10] Charney said. The intruder might also face a misdemeanor charge under Sections 1030(a)(2) and 1030(a)(3) if he accessed, without authorization, certain specified computers (i.e., computers used by financial institutions, card issuers, consumer reporting agencies, or the federal government).

Charney stressed the difficulty of calculating damage, a necessary task in determining whether a federal crime has been committed and how severe the sentence would be. If all the files in Computer B were destroyed, for example, it is not clear how to calculate their value, he noted. Charney also pointed out that, ironically, if the operator of Computer B had complete backup files, any federal crime effectively would be erased. "Does that make sense? I don't think so," the prosecutor observed.   Of course, the victim would most likely incur costs associated with restoring the backup, changing passwords, cleaning up the system, and so on, although the cost of the "ancillary" activities might well be significantly less than the cost of recreating the destroyed files from scratch.

He applauded a recent U.S. Sentencing Commission recommendation for new sentencing guidelines for computer crime that would focus not on economic damage, but rather on confidentiality, integrity, and availability of data.[11] That approach makes sense, Charney said, because in many computer offenses "the real problem is not money, it's privacy and trusted systems, and I think we need to address that."

Guidoboni agreed with Charney that damage calculations are a problem. For example, prosecutors in the *Morris* case claimed substantial amounts of economic damage that would have been dismissed as "speculative" in the average civil case, Guidoboni said.

Issue: Ethics and Education

Charney and Godwin agreed that computer users need ethics edu-

---

[10]This level of damage is likely, assuming that estimates of average losses are accurate. The average loss due to computer crime has been estimated at $44,000 to $10 million (Chen, "Computer Crime and the Computer Fraud and Abuse Act of 1986," 1990).

[11]*Federal Register,* December 31, 1992. As of May 1994, the recommendation (revised since 1992) had not been adopted.

cation.[12] Charney noted that the Justice and Education departments recently published a booklet for educators on the ethical use of information technologies. Many computer hackers are not trying to damage anything; they simply are young and do not know any better, he explained. In such cases, federal agents visit the intruders' homes and tell the parents, and the problem generally stops. "Look, there will always be some percentage of the population that has a criminal bent, and as they learn to use computers, they are going to do nasty things with them," Charney said. "But if you eliminate all these other cases that exist only because the people haven't been educated, then you can apply what limited resources you have left to the really serious cases."

Other forum participants offered conflicting perspectives on the ethical status of computer users. Lance Hoffman, professor of electrical engineering and computer science at George Washington University, felt he was "living in fantasyland, listening to some of this discussion." He suggested that the term "ethically educated computer user" might be an oxymoron, citing a class he taught in which 22 out of 23 students did not feel that stealing information was a crime. On the other hand, Rasch—as noted earlier—suggested that "stealing information" was not a crime.

Still other participants argued that many hackers have a sense of ethics, citing the Legion of Doom as an illustration. The Legion of Doom was a group of several dozen hackers who broke into telephone systems and computer networks. The group acquired widespread notoriety beginning in 1989, when federal agents raided the homes of three Atlanta members who had been trafficking in stolen access codes. One hacker had broken into a BellSouth computer and obtained an administrative document. The crackdown continued into 1990, with raids in Texas and elsewhere.[13]

John Perry Barlow said these hackers "were astonishingly ethical people in spite of the fact that they had been raised in an environment where they had literally been without adult supervision,[14] both

---

[12]The need for ethics education was underscored in the CSTB report *Computers at Risk: Safe Computing in the Information Age* (National Academy Press, Washington, D.C.), published in 1990.

[13]G. Cartwright, "Clash of the Cyberpunks," *Texas Monthly*, Vol. 21, pp. 104-106, 112-116, January 1993.

[14]This may be hyperbole. At least one former member of the Legion of Doom says his parents knew of his activities. Moreover, the ethics of these hackers were self-serving, and they were fully cognizant of the law. For example, members were instructed to avoid causing damage to files—not for altruistic reasons, but so that they could avoid getting caught.

in their homes and in their chosen environment (i.e., the electronic networking environment). They had developed a set of ethics as a natural result of being in an environment where law simply didn't apply." Argued Barlow, "If they didn't have ethics, you would not be able to place a phone call—ever—in this country. They are amazingly careful about what they do. And there is nobody who is in a very good position to stop them." Alexander, who prosecuted the BellSouth case,[15] agreed with Barlow about their technical capabilities, noting that "if they had wanted to shut down service in the entire Atlanta area or up and down the Eastern seaboard or the entire country, they could have."

More recently (mid-1994), public attention has been called to the ethics issue by the National Computer Ethics and Responsibilities Campaign, which aims to provide people with the tools and pointers they need to use information technology in responsible and ethical ways. Sponsored by several different organizations,[16] the campaign is undertaking a variety of activities intended to encourage serious, ongoing discussion of the issue, lend the subject credibility and impact, provide a strong rationale for commentators to focus on the issue, create a structure that gets information into the hands of people who need it, and raise public awareness. Its core message is that users of information technologies have responsibilities that are peculiar to the management, development, and use of those technologies. The campaign is not promoting any approach, "code of conduct," position, or recommendation other than the need to raise awareness of the positive and negative consequences of the analog-to-digital shift and of the fact that tools and resources exist to help people make intelligent, informed choices about how best to develop, manage, and use information technology whether it is in the home, corporation, or classroom.

---

[15]*United States v. Riggs aka The Prophet*, 967 F.2d 561, 1992, U.S. App. LEXIS 17592 (11th Cir. 1992).

[16]Sponsors of the campaign launch activities include the Atterbury Foundation, Boston Computer Society, Business Software Alliance, CompuServe, Computer Ethics Institute, Computer Professionals for Social Responsibility, Computing Technology Industry Association, Electronic Privacy Information Center, Merrill Lynch & Co. Inc., Monsanto, National Computer Security Association, Software Creations BBS, Software Publishers Association, Symantec Corporation, and Ziff-Davis Publishing Company. Affiliated organizations included EDUCOM, the Electronic Messaging Association, the IEEE Computer Society, the California Privacy Rights Clearinghouse, and others.

Issue: Operator Responsibilities and Liabilities

System operators and network service providers have an obligation to provide security capabilities and procedures that discourage
unauthorized access and/or damaging conduct. But users of these
systems and services have obligations as well to use these capabilities and procedures. Guidoboni argued that system operators should
assume some responsibility for security rather than depend on laws
to compensate for bad management. In the *Morris* case, he noted,
passwords had been left in files that were accessible to nearly anyone. In a similar vein, Alexander said failure to install adequate
system security should neither be criminalized nor become a basis
for blocking prosecution of intruders.[17] He argued that common sense
and the free-market system will encourage adoption of safeguards.
More recently, the Clinton administration's Information Infrastructure Task Force has suggested that "users of personal information
must take reasonable steps to prevent the information they have from
being disclosed or altered improperly. Such users should use appropriate managerial and technical controls to protect the confidentiality
and integrity of personal information."[18] Such a principle argues
that individuals have an active responsibility to protect information
about themselves.

COMMON THEMES

An important theme of the dialogue, noted by several participants, was the unusual degree of accord among the factions represented. For instance, a variety of speakers stressed the need for user
ethics education. In addition, prosecutor Charney and civil liberties
advocate Godwin agreed that privacy, more than theft or specific
damage, is the key issue in computer trespassing. (This is not to say,
however, that damage resulting from break-ins is unimportant. Indeed, even when there is no explicit damage, companies may incur
substantial costs if they are forced to take a network out of service

---

[17]In some European countries, failure to install adequate security can nullify any
criminal charges against an intruder; moreover, the system operator can be held liable,
Charney noted.

[18]See p. 3 in *Principles for Providing and Using Personal Information* and the accompa-
nying commentary, distributed for comments by the Information Infrastructure Task
Force Information Policy Committee Working Group on Privacy in the National Information Infrastructure on May 4, 1994. Initial reactions to this document suggest that
this particular principle is the most controversial.

temporarily to check it out and patch security holes. In addition, a user's sense of operating in a trustworthy computing environment may be compromised.)

Perhaps most important, a consensus emerged that federal computer crime statutes—*United States Code Annotated,* Title 18, Section 1030—needs to be updated, although panel members differed as to how. This statute originated with the Counterfeit Access Device and Computer Fraud and Abuse Act of 1984 (Public Law 98-473) and was amended by the Computer Fraud and Abuse Act of 1986 (Public Law 99-474). It provides criminal sanctions for individuals who intentionally access a computer without authorization or exceed authorized access; it does not address reckless or negligent access or intent to cause damage.

The law is difficult to apply in practice. Charney and others pointed out, for example, that insiders pose a significant threat to network security: "We're seeing a fairly dramatic increase in the number of insiders who plant viruses and other types of malicious codes because they're either disgruntled or they're on their way out the door . . . . " These cases are a challenge to prosecutors because the defendants had authorized access to the system. Charney said legislation has been proposed, although not passed, that would criminalize "unauthorized use" instead. Rotenberg also emphasized the threat from insiders, suggesting that the federal statute be revamped altogether to focus on actual harm and criminalizing "people who cause damage and people who intend to cause damage."

Others pointed out that the definition of "authorization" is problematic, especially in a networked environment designed to make it easy to share computing resources. As Rasch said, "We have a question of what is somebody's authorization on a network. . . . As a member of the Internet community, I can send electronic mail to anybody on the network. And I can access their computer, albeit for limited purposes.[19] Therefore, I have authority [under one definition

---

[19]One technical nuance to this point is the fact that Internet e-mail delivery is handled by an independent "postal system" that is not under the direct control of a mail sender. Thus, "Bob's" sending an e-mail message to "Alice" does not give Bob even limited access to Alice's computer any more than Bob's sending a letter gives him access to Alice's postal box. Perhaps the real issue is not who has access (though that is the concern of the law), but rather what is legal to send through e-mail. (For example, what is the electronic equivalent of a mail bomb?) A second technical nuance is that Bob might use another capability provided by most Internet-connected computers, such as "finger," that would in fact require direct access to Alice's computer. (The "finger" command typically provides information about a user with an account on the target computer.)

of the word] to access every computer on the network. . . . Even worse, the problem is not just authorization, it's also exceeding authorization. So even if you are authorized to be on a system, if you exceed that authorization, you may be subject to criminal sanctions." David Hughes pointed out that "it's not just authorization on the Internet, but everything is getting connected to everything else. . . . It looks like we are hooking up every network possible to every network possible. . . . Furthermore, the information is moving, and you don't even know what systems it's going through and whether you have authorization to go through that system." In short, Hughes argued that basing criminality on whether a user has authorized access to a system can raise some perplexing questions.

David R. Johnson, a lawyer with Wilmer, Cutler, and Pickering, pointed out that the Computer Fraud and Abuse Act criminalized unauthorized access only because the Congress was concerned that criminalizing electronic destruction by users with *authorized* access could open up a new avenue for inappropriate prosecution of whistleblowers and others engaged in First Amendment-type activities.

Still, the concordance among the panel members was seen as a positive sign. Anne Wells Branscomb spoke for many when she observed that "the legal profession seems to have made a bridge now to the computer scientists and the user, [and the] managers of some of these systems. . . . [I]n the past I think they have been talking at cross-purposes, have lived in different worlds."

In summary, electronic vandalism can be characterized along at least two separate dimensions. The first involves the nature of the offense: electronic trespass, electronic invasion of privacy, and destruction or theft of electronic property. Electronic trespass can be captured by the idea that a trespasser is executing commands on a computer that he or she does not have authorization to use. Electronic invasion of privacy refers to the unauthorized examination of files in a computer. Electronic destruction of property refers to the unauthorized deletion or alteration of files in a computer.

The second dimension involves the intent or lack thereof on the part of the alleged perpetrator. With a few exceptions, the presence of intent is what determines the extent to which an alleged perpetrator can be prosecuted under criminal statutes; in the absence of intent, liability for harmful actions is generally restricted to civil liability.

Although expressed in terms relevant to computing, these notions are not new. What computing and networking primarily change is the difficulty of determining when a given offense has occurred. For example:

- *Trespass.* The concept of trespass is grounded in physical space that can be precisely delimited. What is the analog of physical space in a medium in which physical space has lost its traditional meaning? Before networking became common, the concept of trespass could be applied to an unauthorized user sitting at a terminal hard-wired to a mainframe. The terminal was located in a physical space, and the user's presence there, if unauthorized, clearly constituted trespass against the owner of that space (who was also usually the owner of the mainframe). But when computers are interconnected all over the world, and the user may be accessing the network from his or her home and thus may not be subject to the jurisdiction of any interested party, on what is the concept of trespass based? Perhaps the unauthorized execution of commands is sufficient, although this interpretation may be problematic when execution is possible without a remote log-in that clearly demarcates a point of (authorized or unauthorized) access.
- *Invasion of privacy.* In a physical space, an invader of privacy may well leave tracks or other evidence that he or she has opened file cabinets or closets. In many electronic environments, it is easy to view files and directories without leaving any trace whatever. Thus, the owners of a computer that has been compromised may have no idea of what information the alleged invader has obtained. (At the same time, other computer systems maintain logs, and so a trace of at least some activities may exist, whereas espionage activity may take place with paper documents, access to which may not leave any observable traces.)
- *Destruction or theft of property.* The concept of electronic property is complicated by the fact that it, like all intellectual property, is intangible. When physical property is stolen, the original owner no longer has use of that property. Likewise, when it is destroyed, the original owner must buy it again to replace it, thus incurring a cost that is comparable to the cost of initial acquisition. But when electronic property is stolen, the original owner may not even be aware of it because he or she still has the use of it. When electronic property is destroyed, and backups are available, it can be replaced at relatively low cost.[20]

---

[20]The status of "low cost" as a relative descriptor must be emphasized. For example, restoring a large database from a backup could require considerable work, although the cost might still be much lower than the cost of reconstructing the database from scratch. Moreover, "restored" files often do not contain the most recent changes made to them (e.g., a file might be restored using the backup made yesterday, but such a backup would obviously not include changes made since yesterday). Even worse, the absence of recent changes might not be noted by the user. Finally, the backups themselves may not be entirely reliable, and so it may not be possible to restore the "backed-up" version of the file properly.

6

Intellectual Property Interests

Protection of intellectual property interests, whether by copyright, trade secret, patent, trademark, export law, or even encryption, is no simple matter even in traditional communications media. Restrictions on photocopying are violated routinely, though certain types of photocopying are protected by law under fair-use provisions. Pop music is created through "sampling" of other compositions, a practice of debatable legality that has parallels in electronic communications. National approaches to property protection differ, while the status of international copyright law is somewhat murky[1]; and some U.S. laws related to intellectual property are so arcane that users may not even be aware of their existence or applicability.

---

[1]International agreements such as the Universal Copyright Convention set minimum standards for copyright laws, but nations still have divergent views of what copyright means, according to Steven J. Metalitz, vice president and general counsel for the Information Industry Association. For example, U.S. tradition holds that "copyright is a limited monopoly that's given to encourage and give incentive to creation," while Europeans view intellectual property as "an extension of the author's personality," Metalitz said. In an effort to promote strong international standards for intellectual property rights, the United States in 1989 signed the Berne Convention for the Protection of Literary and Artistic Works; virtually every major nation is a signatory of that convention. See U.S. Department of Commerce, *Globalization of the Mass Media*, National Telecommunications and Information Administration (NTIA), Special Publication 93-290, U.S. Government Printing Office, Washington, D.C., 1993.

Within this sphere of uncertainty, electronic networks generate special concerns, and indeed the Internet itself is playing a significant role in changing the nature of publication.[2] For example, in the research community, professional societies are investigating ways to offer their journals through the Internet, and some have already begun to do so. Many researchers would like to effectively circumvent the traditional publication process, which can lead to years of delay, and make their work available on the network for free. In other instances, many individuals hesitate to use the Internet at all, fearing—with some arguable justification—that material they created may be appropriated by others without permission.

The research community is not the only one affected. The on-line availability of various publications raises questions regarding the limits of electronic redistribution of print articles in general. In June 1994, the ClariNet Communications Corporation, publisher of an Internet newspaper, was asked by Knight-Ridder Tribune and its Tribune Media Services Division to cease its electronic publication of syndicated columns written by Dave Barry and Mike Royko, due to concerns about too much information piracy occurring on "the net." It seems that a subscriber to ClariNet, the electronic newspaper, sent a copy of a Dave Barry column from the ClariNet, presumably by e-mail, to a nonsubscriber mailing list, where it then reached a Knight-Ridder employee who reported it to executives at Knight-Ridder. In July 1994, the Working Group on Intellectual Property Rights of the administration's Information Infrastructure Task Force released a draft report on how copyright law and practices may need to be updated in an age of highly interconnected electronic networks. The report proposes, for example, that existing law be clarified to ensure that copyright law protects the creator of works that are disseminated through electronic networks.[3]

Although intellectual property is traditionally the domain of copyright, patents, and trade secrets, most of the discussion of intellectual prop-

---

[2]For example, a recent CSTB study, *Realizing the Information Future: The Internet and Beyond* (National Academy Press, Washington, D.C., 1994), found that "there is . . . broad appreciation that a robust market for networked information and resources is fundamental to the success of the evolving National Information Infrastructure." Nevertheless, it is "much less certain . . . how intellectual property protection can or should evolve to fit the networked environment" (p. 160).

[3]See *Intellectual Property and the National Information Infrastructure,* a preliminary draft of the report of the Working Group on Intellectual Property Rights, Information Infrastructure Task Force, July 1994. A revised version will be developed after public comment and input are received.

erty matters in a networked environment is related to copyright, and discussions at the February 1993 forum reflected this weighting. Both scenarios in this chapter are university-based because the university is a domain in which the clash among competing values and rights (e.g., the value of the free and easy availability of knowledge and information versus the right to receive compensation for one's intellectual work) is often played out most clearly.

BACKGROUND: COPYRIGHT PROTECTION

The Copyright Act of 1976 (Public Law 94-553) protects "original works of authorship" that are "fixed in a tangible medium of expression, now known or later developed, from which they can be perceived, reproduced, or otherwise communicated either directly or with the aid of a machine or device."[4] The owner of a copyright has exclusive rights to (1) reproduce the copyrighted work; (2) prepare derivative works; (3) distribute copies to the public by rental, lease, or lending; (4) perform the work; and (5) display the work publicly.[5] These and other provisions are delineated in the *United States Code Annotated,* Title 17, Sections 101-120.

Copyright protection clearly applies to certain types of electronic materials. The 1980 amendments to the Copyright Act of 1976 expressly protect computer programs as literary works. The original expression of information in databases can be copyrighted, though the facts themselves cannot, a position reinforced by the Supreme Court decision in *Feist v. Rural Telephone* (see Chapter 3). This case eliminated copyright protection for white pages telephone directories, and some believe that it has had a significant impact on the business and practices of on-line database providers.

Original and creative text can be copyrighted as well, although enforcing this protection can be onerous. An example raised by attorney Lance Rose in the November 1992 workshop is that long e-mail messages could probably be protected under copyright law, but for practical reasons (i.e., the legal costs) there may be no legal recourse for violations. Other original and creative intellectual creations whose status is ambiguous under current law include the "look and feel" of an on-line service and a short e-mail message created by one person but posted publicly by another.

For users and system operators, it can be difficult to identify

---

[4]U.S. Department of Commerce, *Globalization of the Mass Media,* 1993.
[5]U.S. Department of Commerce, *Globalization of the Mass Media,* 1993.

copyrighted materials, because current legal practice does not require posting of a copyright notice (though notice is still required to qualify for recovery of certain damages). Thus, the burden is on the user to know the law. Nevertheless, at least some network operators, notably commercial information services, strive to keep copyrighted materials off their networks. Commercial software is fairly easy to identify and remove, but graphics are more difficult, according to Stephen M. Case, president of America OnLine. (Even if images are copyrighted, it is relatively simple to strip copyright notices from them in such a way that there is no indication that they are protected by copyright.) Case takes this responsibility to heart, even though he employs contractors to review system files.[6]

A 1993 federal court case, *Playboy Enterprises Inc. v. Frena*, described in Chapter 3, underscores the need for caution, as it held that a finding of copyright infringement did not require an intent to infringe. In another case involving the posting of copyrighted video games on a public system with the knowledge and encouragement of the system operator, a federal court granted a temporary injunction in favor of the copyright holder on the grounds that it was unlikely that a fair-use defense would be successful.[7]

SCENARIO 1: DATABASE AGREEMENT IS VIOLATED

The library of a large university connected to the Internet subscribes to an electronic database (accessible via a gateway on the library system) and an electronic journal (to be distributed to subscribers by electronic mail). The database owner provides access for up to six users at a time, for a flat fee to be paid by the library. A student exceeds the six-user limit on the database.

---

[6]As Case observed, ". . . the reality of it is the customer is a whole [lot] less accountable for anything on the system. They are paying money to us. They are calling us when they have problems . . . . [T]hey are going to hold us accountable. Similarly, from a practical standpoint, any company that believes we are committing an infringing action will hold us liable." Case's point also applies to music as well.

[7]*Sega Enterprises Ltd. et al. v. Maphia et al.,* 1004 U.S. Dist. Ct. for the Northern District of California, March 18, 1994.

Issue: Licenses, Copyrights, and Enforcement Responsibilities

Forum participants generally agreed that the database subscription agreement, or license, should be enforced by the university, not the provider, in Scenario 1. Karen Hunter, vice president and assistant to the chair of Elsevier Science Inc., said her company asks its licensee universities to "reinforce the notion of copyright" and to stress the ethics of database use with faculty and students. Hunter's chief concern is not how but whether the university tries to enforce its agreement. The six-user limit in the scenario presumably applies to a price, she noted; if the university cannot keep the system secure, then the price needs to be renegotiated.

Robert A. Simons, general counsel and secretary for DIALOG Information Services, said universities have a duty to comply with their subscription agreements; the question is whether their controls are reasonable and, if a student somehow bypasses reasonable controls, whether the university has violated the contract. Such an incident probably would not concern the provider, he said, unless it occurred frequently and the university obviously was not exercising any control.

If a faculty member violated the user limit, then the infraction would become more serious, because a faculty member may be viewed as an agent of the university, Simons said. "[For] faculty members, I think there is an even higher duty, not only on their part but particularly on the university's part to educate them as to what is right and what is wrong," he said. "I think the university has an affirmative [duty] to publish 'do's' and 'don'ts' with respect to its network system, much as it has the duty to post a sign on its photocopy machine in order that it not be deemed a vicarious or contributory infringer."

Peter W. Martin, a Cornell Law School professor and former dean, said the university should have a rule and a disciplinary mechanism for dealing with the student in Scenario 1. University rules and regulations traditionally address issues that "get at the core of the academic enterprise" and also affect the university community, and this situation meets those criteria, he said. "[T]he traditional mindset would be to say to any third party . . ., 'That's an internal matter for us; you can't force us to discipline.' And unless a faculty member or library administrator or some other student cares and sort of launches a disciplinary proceeding, it won't happen." At Cornell, Martin said, administrators would care enough about the contractual agreement with the database provider that they would initiate a disciplinary proceeding.

Nancy M. Cline, dean of university libraries at Pennsylvania State

University, was concerned about the need to be responsive to copyright issues and said that expectations of behavior can be conveyed through policy guidelines, faculty contracts, and student codes of conduct; once such expectations are conveyed, the personal responsibility for appropriate use rests with individuals. Beyond that, she was concerned about the prospect of being asked to enforce certain types of contractual provisions having to do with access, pointing out the difficulty of doing so as well as the concern for privacy rights. Her purchasing decisions are based in part on the expectations and obligations tied to particular products. For example, Cline can enforce concurrent user limitations but has difficulty predicting the number of users (a total some providers request to assure profits). It is difficult to "predict safely some of those boundaries without jeopardizing the nature of scholarship," she said.

In that vein, she warned against focusing strictly on tight access controls—whether through strict limitations on the number of users or through the imposition of "per-access" fees that would effectively exclude the majority of students—without considering the purposes of education and research. Universities should not be "blocking off access [to knowledge] and controlling it so rigorously through contract or other types of containment processes . . . that it would change the very nature of scholarship and research. . . . There's a vitality to a lot of research that is predicated on people being able to discover and find and use things that may not necessarily come to them through the usual channels." She continued, "There is a great deal of discovery that goes on in a university context that is serendipitous. People wander into things and thereby branch out and discover a new technique, a new application, whatever. . . . I would hate to see a graduate from microbiology go out into the workplace never having been able to use MEDLINE[8] or some of these other [network] resources just because he or she was the twenty-first student in a 20-workstation contract." She further argued that limitations on access to certain materials reduce the likelihood that those materials will be used and that if people cannot readily use these materials, there may be no need for these materials to be generated in the first place.

Panel members could not offer an ideal mechanism for restricting use. Simons said counting individual users is almost impossible and makes little sense: "Publishers were almost unanimous in agreeing that counting noses or heads is not the same as counting worksta-

---

[8]MEDLINE is an electronic bibliographic database produced by the National Library of Medicine.

tions. . . . It's almost impossible, and it's an impractical approach to try to count noses. It's as if one is saying that without a copyright issue, I'm going to post something with a thumbtack on a bulletin board, but only if you allow 20 people to look at it, and not 21. . . . That doesn't make much sense in either a copyright environment or a commercial environment." On the other hand, the number of workstations with simultaneous access to the relevant files could be restricted. But Ronald L. Plesser, an attorney specializing in new information technologies, questioned whether physical access should be limited. He noted that many universities issue personal identification numbers that can be used at various workstations. He said that pricing strategies for handling such situations will evolve in the marketplace.

Pamela Samuelson, a professor of intellectual property law at the University of Pittsburgh, expressed concern about requiring users to know and abide by all the details of many different licensing agreements. She agreed that universities should foster discussions and establish guidelines regarding appropriate behavior on electronic networks, suggesting that universities help foster a consensus about what is "reasonable within that environment."

Plesser and Simons said current copyright law works fairly well in the electronic environment, though improvements are certainly possible. Plesser said the breakdown comes in enforcement. There are two problems, he said: how to know when someone takes proprietary electronic material and plagiarizes it or alters it, and how to identify the perpetrator so that he or she can be held liable for damages or payment. "We've got to create maybe some new mechanisms of enforcement, collection, [and] payment," he stated.

Emphasizing the distinction between licensing and copyright protections, Plesser said licenses may be issued both for copyrighted works and for databases that may not be copyrighted, and so copyright law is not always relevant to property protection. "The fact that you can't copyright [material] or may not be able to enforce the copyright does not necessarily mean you can't license it, control the use, and control the issue," he said. "So I think the law comes from two places." He noted that the *Feist* case (described in Chapter 3) raised some of these issues.

Simons predicted that a mechanism will be developed whereby users may obtain duplicates of copyrighted electronic materials for a fee; a notice specifying acceptable and unacceptable uses of the materials could be placed on copyrighted material so that users would be sure of the restrictions. A system of this nature would transform electronic networks used for information access into an electronic bookstore, he suggested. "Once we do that, those publishers and authors who are currently worried about placing their valuable data into this morass . . . will say, 'Aha, it's now a marketplace where I can go and buy and sell my wares . . . .'"

Issue: Fair Use

Computer and network technology increases the ease with which electronic text can be copied, distributed, and altered. Thus, David Johnson wondered "if the copyright statutes that arose out of the print mechanisms are themselves becoming hopelessly outdated." He went on to note that we may need to "give up on an effort to protect particular artifacts of electronic text as if they were property and instead focus on how to control access to information through particular channels"; the proper route, he said, was to address "the repetitive distribution of information through channels like libraries and universities and institutions" and what that type of distribution would do to the nature of life in the affected communities.

Licensing agreements for the use of electronic documents alter a university's approach to discipline significantly, by effectively imposing a new definition of fair use,[9] according to Martin. As a condition of providing service, the provider has nearly unlimited power to place any restraints whatever on usage as long as the university agrees to them in advance, he said. Martin described the agreement as overriding fair-use provisions, in that "the leverage is there for the provider to assert through license his own view of what is appropriate distribution of that information, without regard to what fair use under the copyright laws would allow." Martin further argued that "what's new about [the networking environment] and its licensing framework via the universities is that universities cannot, as they

---

[9]Fair use is an exception to the rights of a copyright owner. These provisions (Title 17, Section 107) allow use of a work for limited purposes. Fair use is judged according to (1) the purpose of the use (i.e., commercial or private), (2) the nature of the original work (i.e., fiction or fact), (3) the extent of the portion used (i.e., brief or substantial), and (4) the effect of the use on the economic value of the original work. See M.F. Radcliffe, "Intellectual Property and Multimedia: Legal Issues in the New Media World," *Multimedia 2000: Market Developments, Media Business Impacts and Future Trends,* M. De Sonne, ed., National Association of Broadcasters, Washington, D.C., 1993, pp. 121-148. The recently issued preliminary draft of the report of the administration's working group on intellectual property rights points out that the last of these provisions has repeatedly been held to be the most important in determining the applicability of fair use, a point that is certain to grow in significance as networks make possible much larger markets for such work.

could with print materials, put up signs next to the copy machines [describing appropriate use] and then leave faculty and students to their own judgments, protected by fair-use provisions."

Samuelson, however, said that fair use is determined not by publishers but rather by law. "And I think it's really important to note, especially when you are dealing with university contexts, that there has to be some room for fair use that the community can participate in and not really rely on publishers to define what the scope of use is." Copyrighted electronic materials are covered by the fair-use provisions in copyright law, according to Samuelson; notions of fair use have been extended to the electronic environment in at least two precedent-setting legal cases.[10]

In general, Samuelson said, users are becoming increasingly sensitive to the boundaries of fair use, and a consensus is evolving that caution should be exercised in reproducing and widely distributing the material of other authors without explicit permission. She suggested that copyright law would remain as "a kind of net" but that new mechanisms for handling these issues will evolve as the Internet is used increasingly for commercial services. She went on to predict "a kind of transition from today's environment, in which people basically exchange information without the expectation of the commercialization of that environment, to an environment where commercialization is a more routine [mechanism for distribution]."

Providers indicated that the question of who should determine fair use is difficult to answer. Hunter said publishers are not so much "dictating" fair use as trying to fill a legal void. Part of the problem is that, although the copyright guidelines of the National Commission on New Technological Uses of Copyrighted Works (an advisory commission established in the late 1970s to address copyright issues as they related to new technologies) address fair use in library photocopying, there appears to be no legal precedent for electronic networks. "I think [fair use] will be defined by . . . negotiation and use and a lot of other things, but I don't think it's fair to say that the law provides for fair use in the electronic environment," Hunter

---

[10]Samuelson cited *Galoob Toys v. Nintendo of America* and *Sega v. Accolade*. In the first case, the Ninth Circuit Court of Appeals held in 1992 that Galoob's enhancement of Nintendo's Game Genie video game did not infringe on Nintendo's copyright, on the grounds that the Galoob enhancement did not incorporate any part of the Game Genie software but rather operated on the data bytes generated by that software (thus the enhancement was not a derivative work based on Game Genie) and that the Galoob enhancement did not harm the market for Game Genie (and was thus protected as fair use). *Sega v. Accolade* is described in Chapter 3.

said; "I don't think we have that yet." The confusion also was noted by Plesser, who represents trade associations and private companies.

Issue: Effects of Property Protection Concerns on Authors and Contributors

Some forum participants suggested that concerns about property protection keep potential contributors out of the electronic environment. In running a legal information institute at Cornell that distributes materials on the Internet, Martin finds that many individuals are afraid to contribute because any contribution they make can be appropriated by others: "They consider it like dumping some valuable watches into Central Park—[they] will be gone." He also discerned considerable confusion over the application of copyright law on electronic networks, particularly with regard to how copyrighted information may be used without the author's permission. Plesser agreed that many academics refuse to put their software, texts, or other materials on the Internet, for fear of losing "the value of their creation."

Hunter said the provider community is divided on this issue. "There are a lot of publishers who are very reticent about the whole network environment," she said. "There are others who say it's a reality, it's the future, and we have to get in and test it." (In the period since Hunter made these remarks, such tests have increased in number.[11]) She likened the situation to a story about the designer of a new campus who constructed the building but postponed laying down sidewalks for about six months, to "see what paths people generate. That's the kind of world we're in. I think we're trying to see how . . . [people] want to use the information."

Cline, on the other hand, said some groups of faculty are collaborating on the Internet with great intensity, having established community understandings about network use. Acceptance of the technology has gone so far that some institutions accept electronic publications in tenure evaluations, she said. She acknowledged the legitimacy of concerns about the quality of material "published" electronically, but she did not believe that such publications, suitably juried or reviewed, would be handicapped in comparison to print publications.[12]

---

[11]See, for example, Computer Science and Telecommunications Board, National Research Council, *Realizing the Information Future*, 1994.

[12]Of course, maintaining the standards of review may be problematic if the ease of publication and distribution translates into less care in preparation because of the lower barriers to becoming a publisher or distributor.

SCENARIO 2: FOREIGN USER COPIES
ENCRYPTION SOFTWARE

A university is connected to the Internet. Under a joint effort of its alumni relations and industrial liaison programs, the university also provides library and Internet access for Company X, a small start-up business founded by university alumni, in return for stock options in Company X. To facilitate private communications, the university provides RSA-based public-key encryption software[13] on its host computers, encourages the software's use, and maintains databases that facilitate the lookup of the public-keys of all users using the university as a node. Someone from Iraq accesses the university computers and downloads the software [which is not approved for export to Iraq].

Issue: Obligations to Monitor

Martin, as the owner of an Internet mailbox that may be accessed from foreign locations, said he feels no duty to monitor users who browse through his system. He is not concerned about someone adding or manipulating files because he controls those types of activities.

Other participants felt obliged to monitor their systems. Hunter said providers have some obligation to inspect their materials for violations of law. In the case of journal publication, inspection may be a responsibility of the editor or reviewer of articles to be published. Simons views the export issue as so important that his licensing agreements often require publishers to refrain from using any material that would constitute "encryption data" under the International Traffic in Arms Regulations or U.S. Department of Commerce rules.[14] Otherwise, it would be possible to inadvertently "export"

---

[13]"RSA" refers to a highly secure public-key encryption scheme. Software implementations are available in the United States and elsewhere, although the software is subject to U.S. export control laws.

[14]As listed items on the U.S. Munitions List, certain highly capable cryptographic technologies are regulated by the International Traffic in Arms Regulations (ITAR) administered by the Office of Munitions Control in the Department of State. Specifically, export of such technologies (and directly related technical data) is forbidden. In addition, under the Export Administration Act, the Department of Commerce controls

such data and thus break the law, he said. He suggested that according to certain views of export control, the mere presence of foreign nationals in attendance at an unclassified domestic conference could be considered an export of technical data.

Lance Hoffman wondered whether faculty and universities understood their obligations at all. He cited his experience in teaching a class on encryption that includes students who are foreign nationals; all assignments are done on-line. This combination of factors raises the possibility of inadvertent "export" of materials that could be considered encryption data, which are treated as munitions and barred from export by U.S. law (the International Traffic in Arms Regulations). Hoffman said he doubted that university counsels are familiar with this law.

Issue: Obligations to Turn Over Records

Participants said the university would be obliged to turn over certain records to authorities as specified by law. Alan McDonald, a special assistant to the assistant director of the Technical Services Division of the Federal Bureau of Investigation (FBI), said an FBI request for transaction records would require cooperation from the university, like any other service provider. The situation probably is covered either by the Electronic Communications Privacy Act (ECPA) or the general process of federal criminal procedure; the FBI might obtain a subpoena under *United States Code Annotated*, Title 18, Section 2703, he said.

Plesser agreed that the university would have to comply with the ECPA, assuming that the university, for billing purposes, kept records of which materials were accessed and that law enforcement authorities requested that information. The action required would depend in part on whether the request was for transactional or content data, a distinction that is difficult to make in the scenario, he said. (This issue is discussed further in Chapter 7.)

"The university can't just hand [the data] over on request," Plesser said. "They have to make sure that there is a legal process and that

---

the export of dual-use items, i.e., items that have substantial use that is both military and civilian. Certain computer products with encryption capabilities are regarded as dual-use, notwithstanding a general relaxation of controls on many computer and communications technologies in March 1994. The definition of "export" is subject to some debate.

it's quite defined, and that they are not notifying the person to whom [the information] relates."

COMMON THEMES

The protection of intellectual property proprietary interests is problematic even outside the networking context. Concerns for such protection have become more widespread with the advent of easily available copying technologies such as photocopiers, tape recorders, and video cassette recorders (VCRs). Seen in this light, computer networks are yet another technology to facilitate copying.

At the same time, computer networks have properties that set them apart from other copying technologies. For example, computer networks make easy the large-scale redistribution of copied information. With photocopiers and VCRs, a separate copy must be made for each new recipient. However, on computer networks it can be as easy to send a copy of an electronic document (a book, a musical recording, a photographic reproduction of a painting) to a thousand people as to one person, and sending it to one can be trivial indeed.[15] Moreover, while the laws of physics impose definite limitations on the extent of analog copying, digital copies can be made with perfect fidelity; thus digital copies can be made ad infinitum, vastly increasing the potential scope of redistribution through the remote access capabilities enabled by networks. While a photocopy of a book or a pirated video cassette recording or even a floppy disk must be physically carried from place to place to transfer information, a computer network can transfer information without such restrictions. Thus, laws and regulations based on impeding the movement of physical objects across geographical boundaries are increasingly difficult to enforce in a networked environment.

Previous copying technologies such as tape recorders and photocopiers have resulted in a legal regime based on the concept of fair use; much of the debate today revolves around the extent to which now-traditional concepts of fair use can be sustained in a networked environment. In addition, electronic digital technology changes the economic calculus. Before digital media and the network, the limited number and quality of copies and the limited distribution of those copies made individual copying insufficiently significant economi-

---

[15]Of course, such a statement does not take into account the struggles that many users encounter in their daily travails with user-unfriendly technologies and interfaces.

cally to merit enforcement. Digital reproduction and network distribution enable much broader circulation of higher-quality reproductions at the same expense of money and effort and thus have very different economic consequences and implications.

Finally, digitally stored information can be altered with relative ease. The same technology that enables perfect reproduction also allows changes to be introduced in reproductions with relative ease. A few keystrokes can modify an electronic document; absent the use of technology to guarantee that no changes have been made, such changes are undetectable.[16] For example, digital photographs can be modified to show scenes that never existed, and the viewer will never know that they have been modified. Coupled with the ability to send anonymous or pseudonymous transmissions, the potential to alter information in largely undetectable ways raises new concerns and reinforces old ones about maintaining the integrity of information and protecting the rights of the creators of such information.

---

[16]Technology is now available that can authenticate the integrity of digital data. It is possible to digitally "sign" a given bit stream (file) in such a way that any modification to the file while in transit, on a bulletin board, and so forth can be detected with very high confidence. However, it is also true that digital signature technology is not widely used today.

7

Privacy

Like the topics examined in previous chapters, privacy issues take on unique aspects in the electronic environment. The discussion of Scenario 1 below centers on four issues: the obligation to notify authorities of death threats, general provider practices and responsibilities, liability for violating privacy, and ethical and social issues. The later discussion of Scenario 2 focuses on legal privacy protection for subscriber information, informed consent, and blocking unwanted sales pitches. Both scenarios involve a relatively large commercial or university provider. However, a wide range of sizes and types of providers is possible, and so it is unlikely that one type of analysis or policy will fit all.

SCENARIO 1: OPERATOR LEARNS OF POSSIBLE DEATH THREAT

John and Sally "meet" on a bulletin board provided by a commercial network operator and then begin corresponding through e-mail. The operator has a "standard" set of acceptable use policies that prohibit sending (through e-mail) or posting (on either type of bulletin board) any "defamatory, obscene, threatening, sexually explicit, ethnically offensive, or illegal material."[1] John receives a private e-mail note from Sally's husband saying, "I know what's going on between you

---

[1]This policy is similar to the one used by America OnLine.

ing set up the situation to find out where Sally lives so that he can harm her.

Issue: General Provider Practices and Responsibilities

The mere existence of system rules does not guarantee that the system operator will know the content of messages, as Raymond pointed out. One approach, he noted, would be to remove from the network those users about whom the system operator received complaints from other users. "[I]f there is a complaint that somebody does something that is inappropriate, and you then investigate and find [that] . . . it's not something that requires law enforcement or anything like that, you would then tell that customer that he or she is no longer welcomed on this service."

Raymond further wondered how knowledge of questionable or potentially reportable activities might reach system operators. In the absence of a specific report by a user (as was made in the scenario), he thought that because supervising user activities and message traffic would be difficult and costly, "I would never be monitoring this traffic in any way, shape, or form, and I can't imagine how you would do it," he said. Such a problem also affects network service providers, who carry such large amounts of message traffic that it is for the most part impractical to monitor the contents of all messages transmitted. In such cases, the only practical approach is for the network service provider to establish rules and guidelines on how its services are to be used and to rely on end-user action to bring questionable activities to its attention and/or to monitor public areas of message postings. Once the provider's attention is drawn to a questionable activity, the provider is then in a position to take action (such as to contact the appropriate law enforcement authorities).

Other forum participants expressed various views on how operators should handle these situations. David Hughes of Old Colorado City Communications said he would "communicate with the guy that's making the threat and say, 'You made a threat that I interpret as a death threat. Do you understand the consequences of that?'" Vinton G. Cerf, president of the Internet Society, agreed that operators should not specify responses in advance, explaining that he would "rather say nothing and hold the ethical sense internally, and when a situation arises, judge that myself."

On the other hand, Murray Turoff of the New Jersey Institute of Technology said, "I would like to argue once again for more responsibility on the other side. If you are going to have any policies on what sort of traffic you are going to allow or not allow, you should have clear-cut policies on what you're going to do when those policies are violated."

Issue: Liability for Violating Privacy

A number of participants suggested that the current legal regime, including the ECPA, probably would protect the operator from liability for violation of privacy if he or she notified the police. Harkins argued that Section 2702 (which states that an operator *may* report such a communication to the police but is not obligated to do so) would allow John to dispose of the message as he saw fit without liability; in particular, he could be regarded as consenting explicitly to the operator's disclosing of the message to the police, and both John and the operator would be held blameless for that act. Justice Department prosecutor Scott Charney said Section 2702 states that a service provider who inadvertently learns the contents of a communication that pertains to the commission of a crime may give it to the police; the statute thus may preempt any civil liability for violation of privacy.

The protection might hold even if the threat turned out to be a hoax, according to Rotenberg. If the operator notified police and the husband were arrested, but the note turned out to be a prank played by Sally, then the operator probably still would not be liable, Rotenberg said. He said a good samaritan law[3] would protect individuals who offer help with good intentions but inadvertently cause some harm.

Michael Godwin of the Electronic Frontier Foundation and David Hughes praised Section 2702 of the ECPA. Godwin said that "it strikes a balance between the legal traditions of not imposing an obligation on individuals to prevent crime with benevolence and an allowance for someone who is a good samaritan, who does inadvertently discover that there is crime and wants to disclose it pursuant to the ethical principle just articulated." Hughes called this distinction "extremely important" and desirable.

Issue: Ethical and Social Issues

Oliver Reed Smoot, Jr., an attorney who serves as executive vice

---

[3]Westin described a good samaritan law as a statute essentially stating that "if a passer-by or a bystander intervenes in a situation to attempt to do a good thing, but worsens the situation or complicates the situation, that person will not be responsible unless the bystander happens to be a professional, say a doctor, and doesn't use the professional care that a doctor would bring to intervention in that situation."

president and treasurer of the Computer and Business Equipment Manufacturers Association, said that the threat of murder clearly outweighs privacy interests in Scenario 1, but that a more ambiguous situation (i.e., one that involved a lesser threat) might be more difficult to handle. He said the legal system provides a formal, disinterested process for resolving conflicting values. In particular, Smoot noted that under the criminal statute, "in order for the [operator's] obligation to be created, what has to happen is that a prosecutor has to decide that this is a serious enough problem to get a subpoena, or perhaps convince a magistrate that a warrant is appropriate. And I think maybe that's a better way to handle this, to resolve the conflicting values and to handle the ambiguity, than it is just to say, 'Oh well, the system operator obviously has an ethical obligation to disregard the privacy interest and pass this along,' because that's not at all obvious to me."[4]

Forum participants offered several views of the relationship between ethics and the law. Patrick Sullivan, executive director of the Computer Ethics Institute, suggested that ethics should take precedence in Scenario 1 and that any action taken should be the same as in a nonelectronic environment. "In terms of the ethics of the scenario, there is really no question," he said. "There is an obligation to warn, and one isn't going to split hairs over linguistic interpretations of the message. There will be a duty to prevent harm. . . ."

But Alan Westin of Columbia University contended that legal questions still must be answered. "So you frame [the question] first as an ethical dilemma, but in a litigious society in which numbers of people have said that whoever has the deepest pockets is going to face the lawsuit, and the system operator is the one typically with the deepest pockets in this situation, you can't disconnect the question of what is the legal liability from what is the ethical choice."

Attorney David Johnson suggested that the real issue, whether in an ethical or a legal context, is not what actions are required but rather whether the operator considers the situation carefully. "The ultimate duty may be to simply pay attention to the question and be

---

[4]Similar views have been expressed by Jeffrey Schiller, manager of MIT's campus computer network, as a result of a real situation that occurred at MIT. Campus police feared that a student was committing suicide somewhere and asked the university to search the student's private electronic files for clues as to the location, Schiller said. He recalled that "we decided that potentially saving a life was probably more important than protecting privacy and so it was a pretty clear-cut decision. But I'm sure there are plenty of shades of gray that can get interspersed here, so we need to develop, possibly over time, the precedents of what's appropriate and what's not."

thoughtful about it," Johnson said. ". . . Stone's book[5] on corporate responsibility in a sense makes the point that we can probably, in an organized context like this . . . go a long way by simply having a sense of procedures for deliberating thoughtfully on these questions." Johnson further argued that it would be appropriate for a defendant to assert in court that "even though he made a bad call, he was thoughtful about it."

SCENARIO 2: PROVIDER SELLS USER PROFILES TO MERCHANDISERS

A commercial network operator collects information about the interests and purchases of its users by keeping track of the forums and bulletin boards they use and the purchases they make; it then sells this information to other merchandisers. Users are not asked if they wish to participate in the redistribution of such information.

Issue: Legal Privacy Protection for Subscriber Information

The redistribution of information in Scenario 2 is neither permitted nor forbidden under current law, Harkins said. She asserted that the ECPA allows companies to sell general information about subscribers to electronic services but not the contents of their communications.

By contrast, the law is more protective of the privacy of cable subscribers. The Cable Communications Policy Act of 1984 states that cable operators must notify subscribers annually, in writing, concerning what personal information is collected and how it is used. Subscribers may request that their information not be used for these purposes. George Perry of the Prodigy Services Company noted that, ironically, this provision would cover users of emerging interactive cable services but not users of similar services provided through personal computers and telephones.

Rotenberg advocated applying the cable privacy model to electronic commercial transactions. He argued that intrusion can be greater in the electronic world than outside it, noting that mere participation in a forum or discussion may be disclosed. "The problem is not only

---

[5]Christopher D. Stone, *Where the Law Ends: The Social Control of Corporate Behavior,* Harper and Row, New York, 1973.

that there are commercial transactions which are generated through the use of the network disclosed to others, but that one's participation in a forum or discussion group [can be made available]. The facts of those exchanges or those inquiries may be disclosed to others, and that takes concerns about privacy to a higher level."

Tobin said he would support use of the cable privacy model for electronic networks but not necessarily a legal requirement for such notification. He did stress that apart from legal constraints, companies have a commercial incentive to avoid annoying their customers, who may take their business elsewhere. Providers may use information about their customers so long as the latter know about and approve it, Tobin said. "They may not find out," he added. "And if you want to take that chance as a network operator, then you take that chance; but you could end up with some serious trouble on your hands. . . ." Despite such incentives, however, what types of behavior constitute "annoyance" of customers is the subject of much debate.

Harkins said privacy notice and consent mechanisms are not likely to be legislated for electronic networks because the issues are so complicated and cut across so many jurisdictional lines; five or six senate committees, for instance, deal with privacy statutes. "The instinct, at least [from] what I've seen in the last 5 years, is to push and put a lot of pressure on industry to have its own watchdog system . . .," she said.

Issue: Informed Consent

Participants generally agreed that users should have control over dissemination of their personal information, but they also suggested that this protection may be difficult to assure. Westin asserted that individuals should be allowed to make their own decisions and choices about their personal privacy. He suggested that electronic networks can place unique pressures on privacy, in that electronic networks may be able to compile a "richer, more detailed profile" of a user than can individual companies preparing one-dimensional lists of their customers.

Rotenberg argued that the public should be able to choose in advance whether their personal information may be disclosed, just as they do in everyday life, and that users should set their own risk levels, just as people do in the physical world. "But unfortunately, in the electronic world, the default is in the wrong setting, particularly in this area of electronic communications privacy," Rotenberg said. "The default [setting for] the transactional data is that it may be publicly disclosed, sold, [and] collected unless someone says something to the contrary. So I would pull us back in the other direction and give people a choice about disclosing data [i.e., an opt-in choice[6]]." Providers should assume *all* information is sensitive and therefore provide increased protection, Rotenberg said.

The need to obtain informed consent varies in salience and urgency, depending on the nature of the information and society's attitudes about it, Tobin said. "Opt-in" consent probably is not needed in direct marketing, he said, because the risk of harm is minimal if a provider releases, for example, information about who subscribes to *Time* magazine.[7] But, he argued, "as you move down the spectrum toward medical information, certainly something as sensitive as being HIV-positive or [one's] sexual lifestyle—the potential damage to somebody of that information getting into the wrong hands or even hands that they don't want it in is so great that I think that consent in advance, probably in writing, is required if you are going to do it at all."

Obtaining informed consent is not as easy as might be expected, Tobin said. According to Tobin, at American Express,

> every 6 months there is an attitudinal survey done of customers. We actually tell them what we do with the information, but in a very superficial way. We tell them what it results in. We tell them we take information on what they purchase and merge it with other commercially available information, including some credit bureau information, [and] put their names on mailing lists to offer them products and services in which we believe they may be interested. And so what we try to tell them is what happens. We don't tell them *how* it happens. We also try to elicit from them, in their own words, what they expect, and we find that they basically understand that we take subscription lists. Americans actually have a pretty

---

[6]With "opt-out," certain personal information may be disclosed unless the user explicitly checks a box requesting that the information be kept private. With "opt-in," the information is kept private unless the user explicitly checks a box indicating that the information may be released. Advocates of opt-in suggest that an affirmative action required to disclose otherwise private information is the best guarantor of privacy, but information procurers argue that such a requirement would shrink to relatively small proportions the information they could provide and would damage an industry that has considerable economic and social importance. Opt-in places the burden of obtaining consent on the information procurer, while opt-out places the burden of denying consent on the individual.

[7]Of course, the same lack of sensitivity might not apply to the subscriber list of a newsletter on "paying less tax legally," especially if the Internal Revenue Service were interested in that list to determine targets for audit.

good understanding of what happens in very broad terms. But the
level of sensitivity to it is surprisingly lower than you would expect.

At the same time, Tobin felt that the definitions of "informed"
and "consent" were debatable, and that the potential damage resulting from collection and use of information is difficult to determine.
"We don't want to make that decision [about the use of information]
. . . . We don't want Congress to make that decision. We want the
customer to make that decision," Tobin said. "But the trick here is
how do you explain to these people the complexity of information
that is available to us, what we do with it, and what they get out of
it? . . . I'm convinced that the vast majority of our customers don't
really understand technically what's going on."
Cerf said individuals tend to value convenience over privacy.

> I think, judging from my own behavior and that of others, that we
> have a remarkable ability to ignore risk in favor of convenience.
> And so when you're confronted even with the understanding of
> how much information is floating around about your personal hab-
> its, the convenience of ordering things over the telephone with your
> credit card or sending something on e-mail . . . almost invariably
> overcomes most people's reason for concern about privacy and oth-
> er kinds of risks. And so this raises an interesting question about
> whether we have to contemplate saving people from themselves. I
> have no position at this point, but I raise it as a very interesting
> subject.

Sara Kiesler of Carnegie Mellon University pointed out that "the
way people perceive risks is not always in direct parallel with the
actual risk. That is, people will get very concerned about things like
getting brain tumors from using cellular phones, and they will probably overevaluate the probability of damage. . . . It's difficult for
people to consider negative consequences that they can't actually see
in their minds, and the sharing of information about them is hard to
visualize. [As a result, people do not ask themselves] what bad things
could happen to them as a result of people knowing all these things
about them." This, she said, explains why convenience overrides
risk: people are unable to visualize the negative consequences. Kiesler
also felt this phenomenon explained why people tend to habituate to
warnings (e.g., a message displayed daily on a user's computer stating that electronic mail is not private) and to behave as though those
warnings were never displayed.[8]

---

[8]Kiesler cited the example of transcripts used during the Rodney King case. In this
1992 case, Rodney King was beaten by members of the Los Angeles Police Depart-

The result is what Kiesler called the illusion of privacy. "When you don't have cues around you about who else is there," Kiesler explained, "and especially in electronic forums where you have lurkers [people who only receive messages and never send them] and so on, what happens is that you have an illusion that you are much more private than you really are . . . . [People] continue to say things on networks that they wouldn't say otherwise, even though you warn them." Perry concurred, saying many PRODIGY users are new to computers and do not understand the difference between public bulletin board notes and private e-mail. "And even after you explain it to them, they don't understand. So it really is a very critical problem because the messages look so much alike and users treat them kind of alike." William Dutton of the University of Southern California questioned whether there is any legitimate expectation of privacy on electronic networks yet, in that the courts have ruled out any such expectation on cordless telephones.[9]

Others felt that user education was a reasonable option, even while advocating individual choice about disclosure. Raymond noted that personal communications services will keep track of not only telephone and account numbers but also location data. "Of necessity, additional information will be in that database," he said, "and I think the informed-choice kind of approach—educate consumers as to what goes in there and how it is going to be used—is one reasonable way to handle it, because then they can decide whether it is worth it to have that service or whether they would prefer not having that kind of information collected."

Steven Metalitz, vice president and counsel to the Information Industry Association, argued that the issue went beyond "convenience versus risk" to the broader question of "overall benefit versus risk," posing the possibility that a user might get "cheaper use of a system if he consented or if he did not opt out of your information collection." Metalitz was troubled by "people sitting in this room deciding that the public at large is not sufficiently educated about this. . . .

---

ment. A number of transcripts of police radio conversations describing the incident were introduced as evidence at the trial. Kiesler argued that the police officers in question would never have said the things they did if they had been conscious of the fact that all such conversations were being recorded as a matter of department policy.

[9]Congress decided in 1986 that, based on technological differences, the Electronic Communications Privacy Act would protect conversations on cellular telephones but not those on cordless telephones, according to Steven Metalitz. Court decisions have rejected the claim that government eavesdropping on users of cordless telephone violates the Fourth Amendment. Legislation may change this state of affairs in the future.

[W]e have to avoid the temptation to conclude that other people just can't figure out what their privacy is worth, or . . . that we have to decide for them or put our thumb on the scales for them as to how they should decide these things."

Issue: Blocking Unwanted Sales Pitches

Some forum participants believed that current techniques for blocking unwanted sales pitches, whether in physical or electronic form, are inadequate. For the sake of discussion, Westin suggested the use of an approach used by the Direct Marketing Association (DMA): the DMA maintains a list of consumers who ask to be removed from mailing lists, and each electronic network maintains a master list of users who do not wish their personal information circulated; these lists could be shared among the networks.

A number of speakers, however, felt the DMA system did not apply to blocking sales calls or the sending of junk e-mail. They also noted other problems. Allan Adler noted that use of the DMA list is voluntary and applies only to national—not regional or local—mailers. Tobin said the list is effective only if a mailer uses the exact name and address provided by the consumer. Sullivan noted that consumers who are unaware of the list appear by default to want the mailings, when in fact they may not. Mitchell Kapor said the DMA list is an example of an ineffective private-sector privacy code, saying the association does not enforce the code.

Adler noted that the Congress, recognizing the problems with the DMA list, enacted the Telephone Consumer Protection Act of 1991 (Public Law 102-243), which calls for the Federal Communications Commission (FCC) to establish a method whereby consumers may elude unsolicited telephone sales calls. The sponsors of the law originally wanted to establish a national database of consumers who do not want to receive unsolicited sales calls, but merchandisers opposed this idea. So the FCC decided that every solicitor must keep an in-house "do not call" list, Adler said. If consumers who ask to be placed on the list are called again within a certain time period, they may bring a civil suit, Adler said.

The electronic environment presents both unique problems and singular solutions regarding sales pitches, which forum participants indicated is an increasing problem. On the negative side, Cerf noted that whereas unwanted junk mail may simply be thrown out, unwanted "junk e-mail" could clog an electronic mailbox, blocking messages of higher priority from entering. "On some commercial e-mail services, the size of the mailbox is of some finite length. It runs out after 100 messages and you get this little message that comes back saying there is no room left. If my e-mail runs out of space and the important messages don't make it, I'm going to get annoyed."

On the positive side, several speakers suggested that marketplace mechanisms could resolve these problems. Turoff proposed that merchandisers be required to pay consumers to whom they advertise, with the price set by the recipients. He also suggested that users employ a screening feature to delete material from a particular mailer. David Farber of the University of Pennsylvania, for instance, runs an "invisible script" that quietly discards e-mail from specific individuals and specific bins.

COMMON THEMES

First, the dialogue revealed that many providers and attorneys agreed on several important points, including the efficacy of the ECPA—codified primarily at *United States Code Annotated*, Title 18, Section 2702—and the prudence of allowing users to make their own decisions and choices regarding their personal privacy. Second, it was emphasized that models of ethical practice with regard to privacy can be drawn from networked communities.

On the more general issues about privacy, it is clear that such concerns are not new, and Americans have a strong tradition of wishing to be left alone. At the same time, the right to privacy has never been an absolute one, and its costs and benefits must be balanced against those of public disclosure and/or surveillance.[10] Moreover, electronic networks have a number of characteristics that magnify some of the traditional concerns.

A legal regime has developed around information stored on paper. How this regime should be interpreted when information is stored electronically is problematic. For example, in a technological environment in which the content of a document can be assembled instantaneously from a multitude of sources, what counts as a "record"? What about institutions that deliberately refrain from the collection

---

[10]Westin's recent work with pollster Louis Harris found that the American public divides into three categories on this subject: privacy fundamentalists (about 25 percent of respondents), who are deeply concerned about privacy, want laws protecting their privacy, and want to approve or reject use of their personal information for commercial purposes; the unconcerned (18 to 20 percent), who "couldn't care less about privacy"; and the pragmatic majority (56 to 58 percent), who value privacy but whose definition of unreasonable invasion depends on "whether principles of fair information practices" or "constitutional norms" have been followed.

of certain types of electronic information to forestall requests for such records? Many electronic networks produce vast and unique traces of both verbatim communications and transactional data (e.g., who called whom when, network usage statistics, user credit histories). What traces count as a record?

A second issue is that the many network interconnections give rise to potential conflicts across international boundaries. Different nation-states have different approaches to the protection of privacy (e.g., the United States takes a sectoral case-by-case approach, protecting different types of information in distinct ways, whereas Europeans focus on umbrella rules covering all personal information, according to Metalitz), and there is no international privacy law.[11]

A third issue is that the meaning of the legal term "the reasonable expectation of privacy" (the foundation for U.S. privacy law as enunciated by the Supreme Court in 1967) is not clear in the electronic environment. For example, it is clear that electronic networks tend to encourage frank if not imprudent speech, thus magnifying the confusion over what circumstances provide a "reasonable expectation of privacy." Even today, conversations on cordless telephones tied to a base station are not afforded the same legal protection from eavesdropping as cellular telephones that are truly mobile. The perceptions of the "man on the street" about what is "fair" may conflict with basic assumptions regarding the conduct of business. Matters such as the extent of computer literacy in the public may affect profoundly what constitutes reasonable expectations of privacy.

Given all the dilemmas, a variety of new measures—both technical and legal—likely will be needed to ensure electronic privacy and security. Technology will not be able to provide perfect privacy or security, but technological fixes can provide preventive measures to help reduce the range of risks that are faced by users of electronic networks. Legal measures will be necessary to help deal (remedially, if nothing else) with those circumstances in which it is impractical or undesirable to use the technical measures.

---

[11]Of course, there is little international criminal law, tort law, or intellectual property law either. However, international cooperation on developing common approaches to privacy is not entirely absent. Guidelines for developed nations were issued in 1981 by the Organization for Economic Cooperation and Development (OECD). The OECD guidelines have become the basis for national law in 17 of the 24 member states (which are located in North America, Europe, Japan and Australia), and the guidelines also have been used by new democracies in Eastern Europe, according to Marc Rotenberg. More recently, the European Community (EC) issued a draft data protection directive, which if adopted would ban export of data from the 12 EC member states to countries lacking adequate privacy protection.

8

Common Themes

Computer-mediated communications are now several decades old. In that time, the power of computer technology to change the nature of communication has been amply demonstrated. Early pioneers in the use of computer-mediated communications had some glimmerings that a new medium for discourse was about to emerge, and in recent years an incipient large-scale interest in the use of such communications has proved their forecast to be correct.

At the same time, this large-scale interest has prompted and indeed necessitated serious attention to the social issues that surround the formation and evolution of communities on electronic networks. These issues are complex and difficult to resolve.

VALUES AND NORMS IN NETWORKED ENVIRONMENTS

Networked communities are beginning to grapple with the rules that govern (or that should govern) behavior on electronic networks. Less formal rules of conduct and the means to enforce these rules are emerging as people acquire more and more experience with electronic networks. In many cases, behaviors sanctioned by those who

---

NOTE: The material in this chapter is based primarily on the thoughts of the steering committee, although comments from other workshop and forum participants have been used liberally when appropriate.

have had extensive experience with electronic networks are evolving even as newcomers to the technology are having their first networking experiences. The relevant legal regime is unquestionably changing, as new interpretations of existing laws and even new laws are being enacted, but its presence and potential influence on human behavior on electronic networks cannot be denied. Technology surely has a role in providing tools that help to guide electronic behavior along socially acceptable lines or help to mitigate the consequences of miscreant electronic behavior, but the ultimate issues in this domain are social and political.

Some commentators and analysts believe that the emergence of social norms should be left primarily in the hands of the people who will be affected (i.e., the users of electronic networks). A legal regime (statutes plus the case law that interprets those statutes) that does not make sense when applied to electronic networks will tend to erode the ethical values on which that regime is founded. As a result of such pressures, a set of new values will evolve that will ultimately constitute the basis for a new legal regime. No one is smart enough to take into account all of the ethical issues that will emerge, and so uncertainty is inherent in a situation in which social norms grow and evolve rather than being created de novo.

At the same time, the "natural" evolution of old behaviors into new ones may be problematic and perhaps socially undesirable. A maladapted set of social norms could result for several reasons. One reason is that these rules might evolve in the absence of a real understanding on the part of new users regarding the power and reach of computer-mediated communications (examples are provided in Box 8.1). As George Perry pointed out, "Individuals have never had a megaphone the size of the cyberspace megaphone. Our society has to figure out what to do with this power." Moreover, with new problems come new forms of solutions. Perry recalled the example of the president of a business who complained about something that was on a PRODIGY bulletin board. When told that he could post his own message on the bulletin board to tell people what was really going on, he said, "Oh! You suppose I could do that?" and the problem was solved by that simple action.

A second reason is that electronic networks erase many of the physical barriers—such as geography—to interaction. One important consequence is that electronic networks can bring together people with radically different points of view, moral persuasions, and interests. For example, it is clear that different cultures value different ethical norms, and as a result, different behaviors are considered ethical or unethical depending on the culture. To the extent that people

Box 8.1
Examples of the Scaling-up Problem

On Defamation

A person may write something on an electronic network that is defamatory. Most likely he regards it in the same light as if he had made the remark at a small cocktail party, and he thinks he has every right to defame somebody at a cocktail party. In fact, he does not, but for all practical purposes, he will not be sued for what is said at the cocktail party. Doing the same thing on an electronic network has entirely different social consequences (e.g., he is much more likely to be the target of a lawsuit for defamation), but the lack of immediate cues to those differences may well lead him to behave as though the social environment is the same.

On Copyright

An otherwise law-abiding person may well violate copyright laws in a relatively small way (e.g., by performing excessive photocopying). The electronic networking environment makes it easy to violate copyright laws in a much larger way with approximately the same degree of effort, but again the lack of immediate cues may well lead the person to believe that she is doing the same thing she did with the photocopier.

from different cultures must interact, conflict might be expected. A second consequence is that government and other institutions whose influence is determined largely by the physical control of borders and political boundaries will find that influence challenged by continuing increases in electronic commerce and social discourse.

A third reason is that electronic networks can act to raise barriers with impunity. People who live within the same physical community are subject to important homogenizing influences (e.g., they are exposed to the same newspapers and radio and television broadcasts), and arguably these influences support a common set of values. To the extent that a given electronic community of people is closed to outsiders, traffic within that community can be made immune to external scrutiny and indeed can be entirely insulated from the outside world. Thus, it is not impossible to imagine the existence of different electronic communities that share nothing but the same network protocol.

For these reasons, policymakers are understandably nervous about a legal and social environment that is both fraught with ambiguity and whose future evolution is also uncertain. Thus, the challenge for society at large is to balance the desirability of "natural evolution" against its relatively immediate need for guidelines to mitigate the risks of entirely unrestricted behavior that could affect large numbers of new users. Although even more strenuous outreach efforts to the electronic public and to policymakers will be necessary, an increasing degree of discussion and dialogue between technologists and policymakers in recent years gives rise to the hope that society is beginning to meet this challenge.

RECURRING CONCERNS

Taking a retrospective view of the November 1992 workshop and the February 1993 forum and other events addressing similar issues, it has become clear that several concerns recur consistently. As with many social tensions, an individual is much more likely to attempt to balance competing interests rather than side completely and totally with one of these interests. These points of tension include the following:

• *The extent to which the government should regulate behavior on electronic networks.* For example, a benign view of government would lead people to be less concerned about governmental malfeasance and to be more willing to leave to government the discretion to make appropriate judgments that balance competing concerns. For these people, the law is a way to codify ethical principles and to provide a uniform standard of behavior to ensure order, set social expectations, and provide continuity. A correlate of this position is the belief that society should understand—in advance—how it intends to deal with rare problems that may pose severe difficulties for the individuals affected. Finally, people who believe in an active role for government in these matters are often motivated to find a high degree of certainty in their dealings with the communities involved.

On the other hand, a fundamental distrust of government would make many people unwilling to give government the benefit of any doubt at all; such people would insist that the role of government in regulating behavior be kept to a minimum, and, fearing inappropriate government action, they would oppose government controls or regulations on computer technology and advocate privately negotiated (or community-established) "rules of the road." Such people would believe that in many cases, the law is too heavy-handed and largely incapable of taking into account mitigating or aggravating factors in deciding what should be done in cases of inappropriate behavior, that an absolutely uniform standard is fundamentally undesirable, and that some tolerance of ambiguity is both necessary and desirable.

- *The role of the marketplace in influencing behavior.* The view that marketplace forces can and should regulate and define acceptable behavior on electronic networks would lead people to be less concerned about abuses that may be rare, even if they are severe. These people would assert that large-scale systematic abuse can be controlled by public opinion and pressure and deny the need for laws to regulate their behavior. On the other hand, public opinion and pressure have not entirely eliminated crime and other antisocial behavior in other domains, and so it seems unlikely that the marketplace will entirely prevent computer abuse. An additional dimension of this question is the extent to which social defenses against undesirable behavior should be based on matters of utility or economic feasibility or on matters of fundamental rights.

- *The value of sharing information freely versus keeping information private or proprietary.* The Internet is the preferred medium of communications for much of the scientific community, a community that highly values the free exchange of information. At the same time, the livelihoods of many people depend on their creativity and intellect, and they have an understandable desire to protect the compensation they may receive for their intellectual work. In other cases, individual desires for privacy may conflict with community interests in disclosure. At root, the fundamental difficulty is that as a society, we regard some information as appropriately sharable and other information as appropriately private or proprietary, and there are no clear guidelines for distinguishing between the two.

- *The need for law that specifically relates to behavior on electronic networks.* The assertion that network-specific law is needed represents a claim that the current legal regime is not adequate for application to electronic networks, perhaps as the result of an enabling technology that changes much more rapidly than the relevant legal regime. Nevertheless, it is clear that in the absence of network-specific laws, the application of existing laws to electronic networks will have to be interpreted in light of the new capabilities and limitations of the medium.

- *The extent to which continuity with past precedents is desirable.* Users and providers of network services are, like it or not, inevitably bound by the constraints of the environment in which they operate. But they may have different beliefs about the desirability of such continuity. The belief that continuity is desirable is one aspect of a belief in the value of consistency and social stability. This position would lead people to believe that social continuity, extended to electronic networks, will give users of electronic networks a firmer grounding in what the norms of behavior are (or should be). Those with a weaker belief in the virtues of continuity tend to see electronic networks as an opportunity for designing the ground rules of a new society from scratch and avoiding the socially undesirable difficulties that continuity with the past has created. In addition, they tend to be more concerned about not foreclosing opportunities that the new technology may provide.

- *The nature of informed consent relevant to providing information.* Although all stakeholders endorse in principle the idea that a user of electronic networks making an agreement should give his or her informed consent as a part of that agreement, there is debate over what notions of informed consent should be applicable. Some believe that informed consent should be predicated on understanding the general purposes of the information collector. Others believe that true informed consent requires that all of the possibilities (or perhaps a number of examples that illustrate the true range of possibilities) for the disposition of information be made available to the individual.

A second dimension of the informed consent issue is the disposition of individuals who express no preference or inclination regarding their putative rights on electronic networks. As a general rule, there is a high degree of consensus that the wishes of individuals who explicitly choose to participate or agree or who explicitly decline to participate or agree should be honored. The debate is over the relative propriety of "opt-out" nonagreement (an agreement that applies unless the individual explicitly declines) and "opt-in" agreement (an agreement that applies only if the individual explicitly accepts). There does seem to be something of a consensus that when the potential consequences of an agreement are more severe, opt-in agreement is more desirable. But there is argument over what counts as a severe consequence.

The description of these themes is oversimplified, and it is doubtful that any single individual could be associated solely with any one extreme. It is clear, however, that the positions taken with respect to these themes are reflections of personal value and social ideology or perhaps misunderstandings about how the legal system works, rather than the results of technical deliberation.

In almost none of these situations does electronic networking raise fundamentally new issues. Still, even when old concerns are magnified through use of this technology, dispute and argument become apparent. The reason is that to the extent that these old concerns have been resolved in the past, their resolution has come about not because the concerns have disappeared or the various stakeholders have changed their minds, but because political compromises and the need to move forward have driven decisions. Thus, resolution very much depends on the circumstances of the moment. Networking technology reopens traditional debates largely because it threatens the status quo that results from a given configuration of circumstances; with new circumstances, new compromises become necessary and thus the same fundamental questions need to be reexamined.

If this is true, the debate over social norms on electronic networks, in form and even in structure, will not differ much from the debate over abortion rights, school choice, sex education, crime, welfare reform, or any other controversial social issue. This is not to say that the debates should not be taking place, but only that our hopes about what such debates can accomplish should be moderated. These debates will not resolve fundamental issues or even result in consensus, but they can serve an educational role, illuminating and illustrating issues and providing alternative visions of the future for the concerned public. Ultimately, the debate will be resolved just as all debates over social philosophy are resolved: over time and with a great deal of effort in the courts, newspapers, schools, places of religious worship, and other public forums for argument.

Appendixes

A

Network Technology

NETWORK BUILDING BLOCKS

The essence of an electronic network is connectivity between computers. The first computers ran as stand-alone machines that could be accessed only from their immediate physical location.[1] Later, they became reachable through "dumb" terminals connected to hard-wired lines and dial-in ports. More recently, computer-to-computer connections were implemented through hard-wired and dial-in ports and through local area and wide area networks. Today, many computers participate in the growing global network, sometimes referred to as "the net." One of the main components of this network is the Internet, which itself is a network of networks with international reach. Transmissions over the global network may pass through several computers and network gateways before reaching their final destinations.

When computers are networked, the network must support a method of uniquely addressing the various computers connected to it. Any user regardless of location (but connected to the network) who wishes

---

[1]Even today, there are valid reasons to establish "islands" of computing capability that do not interact with other systems. For example, a corporation may choose not to connect its network or its individual computers to the external world because of security concerns. Still, recent experience demonstrates that important and powerful synergistic effects are possible when many individual computing elements are connected to each other.

to interact with another computer must be able to refer specifically to that computer and not some other computer. Assigning unique addresses to computers on a network is the equivalent of assigning unique telephone numbers to telephones connected to the telephone system.

Computer networks can implement store-and-forward communications, real-time connections, and distributed computing.

STORE-AND-FORWARD COMMUNICATION

With store-and-forward communication, the contents of a communication are temporarily stored on intermediate computers before reaching their final destination. Electronic mail is a good example. A message is typed at the originating computer and is then handed to a "mailer" running on the same computer or on another "mail host" computer tied in through a local area network. Over some period of time, the message will be transferred to and temporarily stored at a series of other computers until it reaches its ultimate destination. At this destination, it is stored on a host computer until the addressee checks her or his electronic mail.

Store-and-forward systems now generally have various enhancements, including the capability to attach files and perform transfers. However, because the native communication protocols of intermediate nodes are not controllable throughout the process, or may in fact be unknown to all systems, file transfers are often limited to text. In any case, the transmission of the message does not create a real-time connection between the sender's and the recipient's computers, and so true real-time interactivity is not possible.

One technical note: packet-switched networks also implement a kind of "store-and-forward" communication of the packets that are the basic unit of transmission. However, intermediate nodes forward packets nearly instantaneously, and these packets remain at the intermediate nodes for very short times.

Real-Time Connections

A real-time connection is one that allows a user on one computer to access a remote computer and directly perform actions on that remote computer. These actions may be as simple as transferring files or searching a database on the remote computer or as complex as controlling the operation of the remote computer. Like the other types of interconnections, real-time connections now often involve several levels of intermediate networks. These intermediate networks are transparent to the user, except that there may be some small time delay induced by the bandwidth, processing, or data communications rate of the intervening physical network media.

Historically, real-time access to remote computers was the first type of networking to be widely developed. It was a practical way around the scarcity of computing power, which was often measured in processor cycles. The idea was to allow researchers to use powerful computers that they did not own themselves. Even with the tremendous decreases in the price of computing power, real-time connections play an important role in providing researchers access to the fastest supercomputers.

Distributed Computing

Distributed computing allows a database, file system, or application to be dispersed across a networked set of computers. Some records in a distributed database or files in a distributed file system may be replicated across several computers to provide greater reliability, faster access, or simultaneous access to a larger number of people. Applications may be distributed in order to take advantage of unused resources on other machines, but they are also distributed so that part of an application can run on a user's workstation while another part runs on a file server or database server and accesses information requested by the user.

NETWORK SERVICES

Networks support many different types of computer-mediated communications. The most basic form of communication, a "one-to-one" message communication between two individuals, is supported through electronic mail and real-time "talk" facilities that allow the parties to simulate a telephone conversation through exchanges of text (advanced systems support voice and video as well). But networking technology greatly expands this basic notion of communication with ease. For example, networks also support the notion of "one-to-many" communication, a mode that could be characterized as a broadcast mode in which a single source transmits information to many people. Perhaps most important, electronic networks support a mode of communication for which there is no close historical analog—a many-to-many mode of communication in which many people write and many people read simultaneously. A many-to-many communications mode that can be operated for essentially the same individual effort as a one-to-one mode is unprecedented in the history of communications media.[2] This mode has facilitated a new pattern of social interaction that is difficult to achieve through other communications media.

Described below are several of the most important forms of computer-mediated communications. Note that these forms do not necessarily map cleanly or uniquely to the one-to-one, one-to-many, or many-to-many modes described above.

Electronic Mail

Electronic mail (or e-mail) is today the single most common form of communication on electronic networks. E-mail has gone mainstream as it is increasingly used in business settings. E-mail is used most often in a one-to-one mode to send private messages from one person to another. However, simply by adding to the address list, it is possible to send (broadcast) the same message to many parties, illustrating the use of e-mail in a one-to-many mode.

The primary advantage of e-mail communication is that it eliminates the need for the message sender and receiver to be active simultaneously; a message sent by one party need not be read by its recipient until it is convenient for the recipient to read it. Postal mail or interdepartmental mail is similar but suffers from the delays and uncertainties inherent in moving physical objects.[3] E-mail is often used instead of the telephone because it solves the problem of "telephone tag." It also provides a written record, and for many people it is free. Still, it is usually more time-consuming to carry on a dialogue using e-mail, and there can be uncertainty about whether the recipient has received a message and acted on requests.

Store-and-Forward Conferencing

Store-and-forward conferencing is a many-to-many mode of communication in which messages are created by members of a group and read by others within that group. Conferencing goes under many

[2]The cost in resources required to support many-to-many communications is an entirely different matter. In fact, the ease with which many-to-many communications can be achieved (from the standpoint of the individual end user) may well mask the true cost in resources needed in support. By all accounts, these resources are substantial.

[3]E-mail is not immune to delays and uncertainties either, although overall delays are generally much smaller than those associated with postal mail. The telephone system, however, is highly reliable and operates in real time. Thus, fax transmission rather than e-mail is often the most reliable way to transmit a single copy of a document.

names (e.g., electronic bulletin boards, "newsgroups," newsletters, forums, mailing lists), but the basic idea is essentially public discussion. People post messages, other people reply, and a structured conversation emerges over a period of time.

The boundaries of the relevant group in a conference may or may not be well defined. Some networked conferences strictly limit those who may participate (e.g., a conference running on a corporate network may be limited strictly to employees of the owning company). Other conferences are essentially public: the group consists of anyone who wishes to join the conference. Still other conferences screen potential members through an application process. In all cases, potential conference members need to know of the conference's existence; in a world in which electronic networks are ever more common, this may be the most daunting admission "requirement" of all.

Some conferences are "moderated"; others are not. In an unmoderated conference, all messages posted by all members are visible to all conference members. In a moderated conference, a moderator screens or reviews messages submitted for posting by group members. Messages that the moderator deems irrelevant or inappropriate for public posting are eliminated or sent back to the sender for revision. For example, a moderator may decide that a message contains a personal attack on another group member and send that message back with a request to rephrase the message.

Properly speaking, an electronic bulletin board is a mode of communication in which all messages ever sent are easily accessible to all participants.[4] A user viewing such a board would be able to see new messages as well as old messages, perhaps dating back to a time before he or she had joined the bulletin board for the first time. However, "bulletin board" has also come to mean an automatic mail redistribution site. In this mode, a mail redistribution site is set up to service discussion on a particular topic among a list of people. If the list is open, then new participants can add themselves to the list automatically by sending a request to the redistribution site. The discussion goes on as users send mail on the topic to the redistribution site, and all members of the list receive it.[5] Old messages—i.e., messages sent

---

[4]Another use of the term "bulletin board" refers to small dial-in computer systems typically operated by private "sysops." This type of system is described below in the section "Computer Bulletin Board Systems."

[5]This characteristic distinguishes it from a simple, individually owned mailing list, in which the sender of a message must know the individual addresses of all those with whom one wishes to communicate. Indeed, in some mail redistribution schemes, it is difficult if not impossible for the message sender to know the individual identities (or addresses) of everyone who receives the message.

among board participants before a given user joins the list—are usually not accessible, unless someone has explicitly archived those messages. (A more proper name of this type of bulletin board is a LISTSERVER.)

Real-Time Conferencing

Real-time conferencing is a many-to-many mode of communication that is similar to store-and-forward conferencing, except that the messages are sent back and forth in real time. Real-time conferencing builds on the real-time connections described in the previous section. One use of real-time conferencing is the real-time chat, in which members of the conference are logged into the conference simultaneously and messages typed by those members are displayed in real time. Real-time chats are the typed network equivalent of citizens' band (CB) radio. Two other forms of real-time conferencing are games such as multiuser dungeons (MUDs) and multiuser simulation environments (MUSEs). MUDs and MUSEs enable remote participants to join ongoing computer-mediated games (dungeons) or simulations. More sophisticated real-time conferences based on audio and video links are beginning to take place on the Internet today.

File Transfer

File transfers provide one type of remote access to information contained in computer files: a user in New York can move a file to or from a computer located in California. A file may contain text, graphic images, or any other type of digitally encoded information. File transfers may be anonymous or restricted. In the case of anonymous file transfers, any user on any computer connected to the network can obtain a file, knowing only the location on which the file is stored and the name of the file. Restricted file transfers are limited to some well-defined group of individuals who, for example, can obtain certain files from certain computers only if they have been granted access rights to those files.

Remote Computer Use

Another way to obtain information remotely is to use the network as essentially a very long cable that connects a user at a terminal or workstation in New York directly to a computer in California—a remote log-in. This user, sitting in New York, has access to capabilities of the California computer that would be available to a user sitting at a terminal that was directly connected to it.[6] Remote use provides a path for the retrieval of information that is not contained in complete files (e.g., the browsing of entries in a catalog, the conducting of a database search) and thus not obtainable through file transfer.

Remote computer use makes it possible to share computing resources; a user in New York in need of a faster computer may be able to find one in California. At the same time, it is more difficult to control remote network-enabled physical access to a computer than to control access when a user must physically appear at a directly connected machine. Passwords and other security measures help to control remote access, but facilitating ease of remote use and denying unauthorized access are goals that are inherently contentious.

Information Search Services

As described above, one major use of electronic networks is to transfer information between sites for the benefit of remote users. But all of the services described above require that the potential user know a lot about the information being sought: on what computer it is located and under what file name it is stored. A potential user that does not know this information can be handicapped in his or her search for information.

A number of tools have been developed for use on the Internet to help users search for and retrieve information (Table A.1). Gopher and the Wide-Area Information Server (WAIS), for example, provide a menu-driven interface for obtaining information from both local and remote systems on the network. Gopher gives the impression of a single large distributed database. Although these and other information search tools are used for different purposes, they share one theme—they reduce the amount of "low-level" information needed by the user to retrieve information, allowing him or her to specify a request for information in terms that are more meaningful (e.g., by a set of key words or phrases that specify the topic of interest on which information is being sought).

---

[6]In many cases, the capabilities available to remote users are restricted due to security considerations. In other cases, the capabilities are identical, and remote users have exactly the same capabilities as do local users.

TABLE A.1 Information-finding Resources for Use on the Internet

Name of Tool	Type of Information Sought	How Search Is Specified
Finger	E-mail address (log-in name) of a user on a given system	Last name of user, system on which he/she may reside
Archie	Usually information in text and binary files	File names to be searched for on various FTP sites
Gopher	Usually information in text and binary files	Menus that contain descriptions of general categories of interest; user browses these menus and is automatically connected to the systems on which menu items of interest are found
Veronica	Files, gopher sites, and menus	Key words to be searched for on various gopher menus, key words associated with various files
Wide-Area Information Server	Information in files	Key words or phrases likely to be found in the files that are desired. (If the file is not text, key words may be appended to auxiliary files that point to the desired nontext file.)
World-Wide Web	Information in files	Hyper-text search; a user browses a document and comes across a reference to locate. (He/she pursues an automatic link from that item in the document to the reference.)
Mosaic[a]	Information in files	Mouse pointing and clicking
Whois	Information about a user without knowing the system on which he/she may reside	Name of user

NOTE: A more complete description of most of these services can be found in Ed Krol, *The Whole Internet: User's Guide and Catalog,* O'Reilly and Associates, Sebastapol, Calif., 1992.

[a]Mosaic is a convenient and easy-to-use graphical user interface to the Internet that became popular in the latter part of 1993 and has been responsible for driving much of the recent Internet use. Mosaic makes use of many of the resources described above.

A ROUGH TYPOLOGY OF NETWORKS

A very rough typology of the various types of networks is the following: global network, computer bulletin board systems, and commercial services. The typology is rough because there is some overlap between the types, but as a first approximation it will suffice.

The Global Network and the Internet

The term "global network" is used to refer to the worldwide connection of computers and networks that is part of and connected to the Internet. The Internet is a set of interconnected networks, numbering in the tens of thousands and most of which make use of a protocol suite known as TCP/IP for communication. The Internet is perhaps the single most important driver of the global network.

Networks other than the Internet have some significance as well, though the networks described here can connect to the Internet. BITNET is a cooperative network consisting mostly of academic institutions and even today provides primary connectivity to network communications for certain institutions. Usenet is a network that supports only newsgroups. UUCP refers to an associated network that supports only mail but not newsgroups. Fidonet is a network that connects personal computers, primarily those using MS-DOS.[7]

Computer Bulletin Board Systems

From a technical perspective, a bulletin board system (BBS) is quite simple. In its most basic form, a BBS involves a computer with a modem on a telephone line. BBS users make their connections by telephone, and data flows between the user and the BBS. Connections between user and BBS are transient, lasting only for the duration of the telephone call. A user dials a connection to the BBS and may interactively read "messages" posted by others, and, if authorized, post his or her own messages. The connection from the user to the BBS is typically a terminal-style interface, e.g., a VT100 terminal emulation, rather than e-mail in the sense described above.

BBSs are an example of grass-roots computing—inexpensive, generally informal, open to everyone with a modem and a computer, and often

---

[7]For more discussion of these and other networks, see John Quarterman, *The Matrix: Computer Networks and Conferencing Systems Worldwide*, Digital Press, Bedford, Mass., 1990.

short-lived. Some BBSs charge for usage, others seek voluntary contributions, and still others are entirely free of charge. BBSs cover a wide range of subject material, e.g., coin collecting, parakeet raising, politics, lifestyle, religion, law enforcement, distribution of government information, and hacker information.

The low cost of setting up and running a BBS has allowed many individuals to establish their own. In 1992, Jack Rickard, editor of *Boardwatch* magazine, estimated the number of publicly accessible bulletin boards in the United States at 45,000, and the number has grown substantially since then.[8] Many BBSs are connected to each other or to other networks; other BBSs stand alone. Freenets, of which one of the most famous is the Cleveland FreeNet, are community-based networks that are open to the public and provide BBS capabilities. They serve many of the same functions as public libraries and town meetings.

Commercial Services

A number of commercial networks have emerged in the last decade. Among the most prominent are CompuServe, the Prodigy Services Company, GEnie, and America OnLine. Although the services provided by commercial networks vary, they generally include access to a variety of information sources (e.g., magazine and newspaper articles, financial information for publicly owned companies), electronic mail among subscribers, public conferences on a variety of subjects ranging from romance to nuclear energy, and a variety of consumer services (e.g., home shopping). In general, commercial services charge users for the time they are connected to the network and for the specific services they use (though a set of basic services may be available for a fixed fee). Policies exist regarding acceptable use of the network services offered, and they are enforced to varying degrees.[9] Common carriers (e.g., local exchange companies) may begin to offer similar services in the near future.

---

[8]This total includes only systems run by either individuals or companies that would welcome a call from a stranger. These bulletin boards typically host between 200 and 2,000 callers each, with an average "unique" caller base of about 250 per board (discounting the common caller base among boards), according to Rickard. This indicates a total U.S. caller base of over 11 million

[9]For example, a commercial information service might offer a public "chat" service to its users (i.e., a conference), subject to the condition that users not engage in the use of profanity. These "terms of service" may stipulate that participants using profane language are subject to disconnection from the conference, but in actual practice action may be taken only when another participant complains. Another service might terminate the connection when the profane language first appears.

THE INTERNET

The Internet is a worldwide network of networks that originated in research and education communities but now also accommodates some commercial traffic. Member networks share a common set of protocols that enable communication between them, but each member network is administratively distinct in much the same way that a given road might pass through a number of separate and distinct states with different law enforcement practices and rules of the road. As a result, the Internet does not have the character of a single, centrally run organization, though certain aspects of its operations are coordinated. The Internet provides all of the functions and services described in the section "Network Services," above.

Organization

The organization of the Internet is best described in terms of progressively higher levels of aggregation:

- *The local view.* Typically, the local view centers on an institution. The institution has a mainframe computer or a local area network, and people use terminals, personal computers, or workstations that are physically wired to the network or the mainframe.
- *The regional view.* At the next level of aggregation, each of those local networks or mainframes is connected into some sort of a regional network. Thus, the regional network consists of many interconnected state or local networks. For example, the California Education and Research Foundation Network (CERFNet) alone connects more than 150 academic and commercial institutions to the Internet. There are a few dozen regional networks, most of which began as academic cooperatives and some of which have become commercial enterprises. In addition, several fully commercial enterprises now provide Internet connections to institutions.
- *The national view.* Regional networks serving research and education users have been connected through a high-speed "backbone" network known as NSFNET, which has been run with financial support from the National Science Foundation since 1988. (NSFNET is expected to be replaced in 1995 by a web of competing commercial and nonprofit backbone networks.) Some networks outside the United States are multinational, e.g., NORDUNET (Scandinavian nations) and DANTE and EBONE (Europe).
- *The global view.* The Internet now reaches internationally. In recent years, various links have been established to networks in other

countries. However, the detailed structure of networks in these other countries may not parallel that in the United States. (For example, networks internal to other nations are often similar to regional networks in the United States.)

But even these views do not quite explain the "true" nature of the interconnections among Internet institutions. For example, multiple connections between regional mid-level networks on an ad hoc basis are common; it is even possible for individuals to become Internet sites on their own. The resulting network is more like a seamless and tangled web of interconnections that conform to the common protocols than a hierarchically structured organism. Among other things, the Internet's lack of hierarchy makes it far more robust and adaptable to new circumstances.

Management

The management of the Internet is decentralized. All Internet sites share communications protocols and an agreed-upon set of naming and addressing conventions. A central body, the Internet Assigned Numbers Authority, allocates Internet addresses. Day-to-day operation is conducted by a set of hierarchically related Network Information Centers. Other than these shared elements, very little is common, though a spirit of friendly cooperation enables this decentralized operation to function. As Susan Estrada, former executive director of the California Education and Research Foundation Network, noted, "everybody or nobody" runs the Internet.

In the past, the decentralized character of the Internet has interfered with enforcement of the NSF acceptable use policy (AUP). The AUP attempted to regulate the nature of traffic carried across NSFNET and was intended to restrict the use of NSFNET to research and education purposes. However, many commercial enterprises are (and have been) connected to the Internet. Since a user generally has no way to control the precise routing of any given traffic, a commercial user may send commercial traffic across NSFNET without knowing it (or even caring about it). Indeed, the NSF AUP has often been honored more in the breach than in actual practice. (It is expected that efforts to apply this policy will change in 1994-1995 as NSF support for the backbone is reduced and the backbone network is privatized.)

Size and Scale

Since it is so easy to add a connection to the Internet, the size of the Internet changes rapidly. The Internet connects more than 70 countries (around 150 countries if e-mail links to the Internet are included), between 2 million and 20 million users, and some 3,000 newsgroups. It also connects many thousands of information archives of various sorts. The Internet connected 46,000 domains in July 1994,[10] and the number of added networks doubles every year.[11] These numbers are growing rapidly. For several years, the traffic across the NSFNET backbone (measured in terms of number of packets of information carried) has grown at an average rate of 15 to 20 percent per month, driven primarily by increases in the number of users.

A BRIEF HISTORICAL BACKGROUND

The earliest roots of computer networking can be clearly traced back to the time-sharing services of the 1960s. At that time, computers were large and expensive, and consequently rather scarce. Time-sharing, using remote terminal communications, was developed as a way to expand the availability of limited computing resources. General Electric and Tymshare were among the better known, early commercial time-sharing services; many universities and companies maintained their own time-sharing mainframe computers. As computers became more common, many corporations established their own mainframe-based local area networks, often with proprietary networking technologies. Subsequently, they linked a number of their own computing sites to form the first long-haul or wide area networks. Xerox, Digital Equipment Corporation, International Business Machines (IBM), and American Telephone and Telegraph (AT&T) were among the early pioneers. Also beginning in the late 1960s, networking began to move out of the single-corporation realm with the establishment of several small, experimental, packet-switched networks in Europe to link scientific research facilities.

Today's widespread availability of fast, reliable, global, and nearly ubiquitous networking is directly related to two significant develop-

---

[10]A "domain" is approximately one administrative entity that is connected to the Internet; typically, it is a single university or a single company. The domain may be subdivided into a number of smaller subdomains.

[11]Vinton Cerf, personal communication, 1994.

ments: ARPANET and personal computers. ARPANET was developed in 1968 by the Advanced Research Projects Agency of the U.S. Department of Defense and implemented in 1969 by Bolt, Beranek, and Newman both as a network research project per se and, as it turned out, a very successful method to link military research computers. It demonstrated the viability and system-wide reliability of long-haul packet-switched networks. The development of the Transmission Control Protocol (TCP) and the Internet Protocol (IP) in the mid to late 1970s enabled the linking of a growing number of wide area and local area networks via ARPANET and thus greatly increased the number of researchers with network access. This linking of a number of networks eventually led to the use of the name "ARPA Internet" in 1977 to stress the internetwork aspects of this growing resource for scientific research. More formally, "Internet" was formed in 1983, when the Defense Communications Agency reorganized military networking and mandated the use of the TCP/IP protocols for all hosts on military networks.

In the mid to late 1980s the National Science Foundation (NSF) established a number of supercomputer centers to make greatly increased computing power available to the broad spectrum of research scientists outside of the military research community. After some initial experience using ARPANET, NSF established the NSFNET backbone for the Internet in 1987 and 1988 and began to link an increasing number of colleges and universities to the network. This greatly increased both the capacity and number of users on the network and reinforced the fact that the original ARPANET had become only one of the many networks on an already large and continually growing Internet.

B
Workshop Schedule and Session Descriptions

SCHEDULE

Thursday, November 5, 1992

8:00 a.m.	Continental Breakfast
8:30	Welcome
8:45	Session 1—Internet
10:30	Break
10:45 a.m.	Session 2—Commercial Information Services
12:30 p.m.	Lunch
1:30	Session 3—Grass-roots Networks
3:15	Break
3:30	Session 4—Mapping Different Network Services Onto Different Metaphors
5:30 p.m.	Reception and Buffet Dinner

Friday, November 6, 1992

8:15 a.m.	Continental Breakfast
9:00 a.m.	Session 5—Content, Censorship, Accuracy, Defamation
Noon	Lunch
1:00 p.m.	Session 6—Privacy and Proprietary Interests
4:15 p.m.	Session 7—Summary and Wrap-up

SESSION DESCRIPTIONS

The first three sessions are to be devoted to examining user, provider, and outsider perspectives on different types of networked communities. Sessions 1 to 3 will address the following questions:

• What policies, laws, regulations, or ethical standards apply to the use of these services, who sets them, how are they developed, and how are they enforced?

• What are users' expectations regarding privacy and protection of other proprietary interests?

• What are the rights, responsibilities, and liabilities of providers or operators of these services?

• What are the rights, responsibilities, and liabilities of users of these services?

• What problems arise from connecting systems offering these services to systems that operate under different policies?

Sessions 5 and 6 are to be devoted to examining important issues that cut across different networked communities.

Session 1—Internet

Chair: Stephen Kent
Presenters:
> Jeffrey Schiller (Massachusetts Institute of Technology)
> Susan Estrada (FARNET)
> David Farber (University of Pennsylvania)

The Internet is the largest network in the world, connecting over a million users through thousands of sub-networks running through universities, industry, government agencies, and other organizations. This session will identify the rights and responsibilities of the Internet community, including the organizations that offer Internet nodes or gateways and the people who use electronic mail and other Internet services.

Session 2—Commercial Information Services

Chair: George Perry
Presenters:
> Stephen Case (America OnLine)
> Murray Turoff (New Jersey Institute of Technology)
> Patrick Lanthier (Pacific Bell)
> Eberhard Wunderlich (AT&T)

Commercial information services (including enhanced services to be offered by common carriers) offer (or will offer) the general public on-line information, bulletin boards, electronic mail, and conferencing facilities, and consumer services to their subscribers. This session will focus on the rights and responsibilities of the providers of such services and members of the public who use them.

Session 3—Grass-roots Networks

Chair: Mitchell Kapor
Presenters:
      Tom Grundner (Cleveland FreeNet)
      Jack Rickard (*Boardwatch*)
      William Dutton (University of Southern California)

Grass-roots networks encompass efforts like FreeNet and local community bulletin board systems such as those operated by Santa Monica or by libraries; they are characterized as being generally small in scale (compared to the Internet) and have a populist flavor.

Session 4—Mapping Different Network Services Onto Different Metaphors

Chair: Anne Wells Branscomb
Presenters:
      Davis Foulger (IBM)
      David Johnson (Wilmer, Cutler, and Pickering)
      Henry Perritt (Villanova University)

A number of metaphors (e.g., printing presses, corner soapboxes, telephones) have been used to describe electronically networked communications. But all such metaphors break down at some point. This session addresses what is special about electronic communication and how the metaphors generally used to understand electronic communication succeed and fail.

Session 5—Content, Censorship, Accuracy, Defamation

Chair: George Perry
Presenters:
      Sara Kiesler (Carnegie Mellon University)
      Carl Kadie (University of Illinois, Urbana-Champaign)
      Allan Adler (Cohn and Marks)
      Jean Polly (NYSERNet)

Questions to be addressed in this session:

- Given the global but largely selective reach of networks, what characterizes the judgments of a given networked community regarding content?
- On what basis are policies regarding acceptable use formulated?
- What policies, laws, and ethical standards establish acceptable content?
- What is the impact on various stakeholders of violating acceptable use policies?
- What is—and should be—the operator's responsibility for user violations of acceptable use policies?

Session 6—Privacy and Proprietary Interests

Chair: Dorothy Denning
Presenters:
William A. Bayse (Federal Bureau of Investigation)
Steven Metalitz (Information Industry Association)
Lance Rose (Attorney)
Alan Westin (Columbia University)
Marc Rotenberg (Computer Professionals for Social Responsibility)

Questions to be addressed in this session:

- What is the nature of the proprietary interests held by the various stakeholders in networked communities?
- What considerations should be taken into account to determine the legitimacy of asserted proprietary interests?
- How can these proprietary interests be protected?
- How have different networked communities acted to safeguard or deny these interests?
- What are the responsibilities of a service provider to protect the privacy and proprietary interests of the users?
- What should be the obligations of providers (or communities) to assist law enforcement or other officials (with or without court orders) in providing access to electronic communications (contents, addresses, and so on)?
- What government controls, if any, should be placed on the use of cryptography?

Session 7—Summary and Wrap-Up

Since this workshop will provide the intellectual underpinning for the next event (i.e., the forum in Spring 1993), it will be important to summarize what the group learned during these two days and what types of questions would be relevant for the forum. The committee chair will summarize key points and then invite discussion from committee members and other workshop participants.

APPENDIX

C
Forum Statement of
Purpose and Agenda

STATEMENT OF PURPOSE

Participation in electronically networked communities is growing by leaps and bounds. The environments for networking include the Internet, commercial network service providers, local bulletin boards, and both intra- and inter-enterprise networks. Network-based businesses are proliferating and growing, and non-profit networks, particularly those that serve the research and education communities, are rapidly expanding their services.

Growth in electronic networking raises many policy issues. How much, if at all, can network service providers restrict access to or specific uses of their services? How much, if any, responsibility do network service providers have to safeguard the privacy or proprietary interests of their users? How do the expectations of individual and corporate network users accord with existing laws? The responses of the general public, private organizations, and government to these questions will shape the progress and impact of electronic networking in U.S. society.

Building on its historic concern for nationwide information infrastructure, the Computer Science and Telecommunications Board (CSTB) chose to focus its second strategic forum on key policy issues associated with the conduct of electronic networking activities. It appointed a steering committee chaired by Dr. Dorothy Denning to organize the project's activities. As a first step, CSTB hosted a small workshop in November 1992. Prominent researchers and policy analysts were invited to air their views in a roundtable discussion on a variety of questions concerning rights and responsibilities in networked communities.

The present strategic forum builds on the November workshop discussions and expands the dialogue to a wider community. To achieve this broader discussion, forum panelists and moderators will explore several scenarios that illustrate the kinds of questions, issues, and choices that must be made in operating, managing, and setting policies for networked communities. The scenarios for panel discussions are described at the end of this program.

AGENDA

Thursday, February 18, 1993

5:00 p.m.	Registration and Reception
6:00	Dinner
7:00 p.m.	Welcome to the Academy by Philip M. Smith, Executive Officer
	Keynote Speech by Congressman Edward Markey, Chair, House Subcommittee on Telecommunications and Finance

Friday, February 19, 1993

7:30 a.m.	Registration and Continental Breakfast
8:30	Welcoming Remarks
	• Frank Press, National Research Council Chair
	• Dorothy Denning (Georgetown University), *Forum Chair*
8:45	Setting the Stage
	• Technical Tour Through Cyberspace— Mitchell Kapor (ON Technology)
	• The Legal Landscape of Cyberspace— Anne Wells Branscomb (Harvard University)
	• The Legislative and Policy Context of Cyberspace— Richard Wiley (Wiley, Rein, and Fielding)
10:15	Break

10:35   Free Speech
                • Henry Perritt (Villanova University School of Law),
                  Moderator
                • Allan Adler (Cohen and Marks)
                • George Perry (Prodigy Services Company)
                • Reid Crawford (Iowa State University)
                • David Hughes (Old Colorado City Communications)
                • Lawrence Lessig (University of Chicago)

11:45 a.m.   Lunch with Special Presentation by John Perry Barlow
                (founder, Electronic Frontier Foundation)

1:00 p.m.   Electronic Vandalism
                • Oliver Smoot, Jr. (Computer and Business
                  Manufacturers Association), *Moderator*
                • Kent Alexander (King and Spalding)
                • Scott Charney (Department of Justice)
                • Michael Godwin (Electronic Frontier Foundation)
                • Thomas Guidoboni (Bonner and O'Connell)
                • Mark Rasch (Arent, Fox, Kintner, Plotkin, and Kahn)

2:00   Protection of Proprietary Interests
                • David Johnson (Wilmer, Cutler, and Pickering),
                  Moderator
                • Nancy Cline (Pennsylvania State University)
                • Karen Hunter (Elsevier Science Publishing Company)
                • Peter Martin (Cornell Law School)
                • Alan McDonald (Federal Bureau of Investigation)
                • Ronald Plesser (Piper and Marbury)
                • Pamela Samuelson (University of Pittsburgh)
                • Robert Simons (DIALOG Information Services)

3:00   Break

3:20   Privacy
                • Alan Westin (Columbia University), *Moderator*
                • Ann Harkins (Senate Judiciary Subcommittee on
                  Technology and the Law
                • Kenneth Raymond (NYNEX Telesector Resources
                  Group)
                • Marc Rotenberg (Computer Professionals for
                  Social Responsibility)
                • James Tobin (American Express Company)

4:30   Synthesis and Concluding Remarks by the
Forum Steering Committee
- Dorothy Denning (Georgetown University), *Forum Chair*
- Anne Wells Branscomb (Harvard University)
- Mitchell Kapor (ON Technology)
- Stephen Kent (Bolt, Beranek, and Newman)
- George Perry (Prodigy Services Company)

5:00 p.m.   Adjourn

D

Keynote Speech:
Networked Communities and the
Laws of Cyberspace

Edward Markey
Chairman, House Subcommittee on
Telecommunications and Finance

As many of you are aware, we are in a time of transition. As of last night, the Clinton administration has begun the political transition from 12 years of Republican economic policies, launching the nationally televised "airwave" assault of Operation Shared Sacrifice with an address to Congress. The "ground war," wherein we get into the nitty, gritty details of legislation, will come later this spring.

Although it is an ancient Chinese curse to live in interesting times, it's an exciting time to be here in Washington because of all the transition and the change.

Everyone is throwing out the old speeches about gridlock and divided government, the misguided regulatory state, and the lack of vision. And you should see what the Democrats are doing. New material is being written all over town. Gone are the jokes about George Bush and grocery scanners. Tough to get a smile with a line about Dan Quayle or anything these days. If he were still around, somebody could crack a joke about Vice President Quayle, as head of the Space Council, proposing that the country build a "cyberspace

---

NOTE: Speech delivered on February 18, 1993, for the CSTB strategic forum. The version here was transcribed from a tape of the speech, supplemented with the printed text of Mr. Markey's speech.

station" before the Japanese do. But he's not around, so we won't use any of those jokes.

It is a good time, instead, to talk of technological transition and the change it is spawning throughout industries and indeed, throughout American society. As all of you know, technologies are moving increasingly from analog to digital forms of communication, and whole industries are undergoing a transition being wrought by the convergence, or growing together, of digital technologies.

This technological transition or convergence is leading inexorably toward the creation of a new mega-industry: the information industry. Comprising computer companies, software houses, telephone and cable companies, manufacturers of wireless gadgets, and others, the development of this mega-industry and its harmonization of what we more often think of as diverse technologies and distinct industries are taking place at a heady pace.

A quick look around at what is happening out there can astound even those who follow such developments, as I do, rather closely. For instance:

- Hollywood is going digital. New computer-controlled special effects like "morphing," which allowed villainous characters in *Terminator 2* to constantly change form, digitized movies, and computer-simulated prescreening, which saves producers thousands of dollars, are becoming commonplace. The "high-tech" efforts of today are allowing "digital actors" from yesteryear, like Humphrey Bogart or Groucho Marx, to come alive again on the screen, or to dance with Paula Abdul in soft-drink commercials. Columbia has talked about releasing its movies and interactive video using digital distribution over telephone lines and cable TV to homes and theaters.
- In health care, a recent study by Arthur D. Little concluded that advanced telecommunications could help cut the cost of health care delivery in this country by $36 billion annually through "telemedicine," remote video doctor-patient consultations, and electronic filing of claims.
- In education, the advent of digital communications in the classroom, from ISDN-based distance learning projects to interactive media as a new tool in teaching, could help make the telecommunications infrastructure of tomorrow the "great equalizer" in U.S. education by linking resources and students, rich and poor, urban and rural, and by giving everyone access to the Information Age.
- In manufacturing, the agile techniques that telecommunications-conversant CEOs employ could mean that the "virtual corporation" may help Bill Clinton achieve a "virtual economic recovery."

In short, the rapid technological change and the new networks being created will increase efficiency and create thousands of jobs.

All of this excites the imagination. And I'm excited to have a "front row seat" to it all as chairman of the House Subcommittee on Telecommunications and Finance, which oversees and legislates in the area of telecommunications. At its heart, it's the subcommittee of MTV and Hollywood, of telephone networks, cable companies, and computers.

Historically, though, policymakers and regulators have tended to look at the telephone industry, the cable industry, or the television and broadcast industries as distinct entities defined by their technologies. We speak of wires and switches, antennas and cables, regulated services and tariffs. Often, policy debates revolve around intraindustry or interindustry tugs of war, with legislators and regulators feeling a little like the mythical creature of *Dr. Doolittle*—the "pushme-pullyou"— a llama with two heads facing in opposite directions. Policy stasis was frequently the result. In such scenarios, creating a national network and articulating overarching policy goals for the country can be an arduous task.

I do feel, however, that this year we are on the cusp of dramatic change that will bring success. With Bill Clinton and Al Gore, along with Ron Brown at the Commerce Department, we have an administration that understands the nature of technology and the economic role it can play. Moreover, they are also people who understand the "small-D" democratic potential of these new information networks and place more importance, I believe, on some of the basic principles of personal privacy that we should all be vigilant in protecting. We've gone from a president who didn't know about grocery scanners to one who not only knows what a "PBX" is, but knows the capacity of the one in the White House and is unhappy with it!

Tonight's forum, where we talk of the nature of "community" in a networked world and the rights and responsibilities of the members of that community, gets us beyond the wires and the switches to the electronic culture that exists on the ends of the line and between the wires—the amorphous, borderless world of "cyberspace." Although cyberspace existed before author William Gibson coined the term in his book *Neuromancer*, word of this new realm, this electronic frontier, has only begun to capture the imagination and attention of the general public recently.

Of course, appearing on the cover of *Time* magazine helps to get the word out. (So, "cyberspace" does have newfound name recognition—although it would probably have to make some rounds at Manchester coffeehouses before it would be a top vote-getter in the New Hampshire primary.)

What happens in cyberspace, and what its popular emergence portends for our society, is what I'd like to talk about tonight. I want to step back and highlight some of the tough issues that lurk just over the horizon of our electronic frontier. It is important to raise the issues now. This is a new arena for communications and law. Conferences like this one provide a great service by sharing information and perspectives and by trouble-shooting potential problems. In short, when we begin to address the rights and responsibilities of cyberspace, we will be boldly going where no policymakers have gone before.

Before we can discuss "communities" in a networked world— before we get to the larger, national network—we must first look at an individual human being and that individual's relationship to the technology.

When Harvard's Mark 1 computer and the Eniac computer were being brought on-line back in 1944 and 1945, the budding dream of many a computer scientist was to create ever more complex machines. The ultimate goal was to create a machine that could think like a human being, to move beyond the simple logic, the programmed syllogisms, the endless zeros and ones—to artificial intelligence and thought processes that would mimic and indeed rival humans.

Science fiction novelists have long written of such machines and robots. The ambition was to have machines play an integral role in human society. The paradigm has changed in recent years. Ironically, today we talk of what role humans can play in an interconnected network of machines.

The culture that develops in cyberspace will have both good and bad elements, analogous perhaps to the "thinking machines" dreamt of in science fiction, which sometimes acquired human qualities we ourselves would often like to forget. Think of "Hal," the on-board computer that took over control in the movie *2001: A Space Odyssey*, and one is reminded that there can be less than benevolent consequences from technological advances.

Ironically, after World War II, as computers began to evolve and become more sophisticated, a debate was also raging in the field of human psychology—the behavioralist school was maintaining that humans were actually more "computer-like," in the sense that human actions and behavior stemmed from predictable logic or conditioned behavior. Hungarian writer Arthur Koestler, more famous perhaps for writing the antitotalitarian classic *Darkness at Noon* in 1941, wrote another novel, *Ghost in the Machine*, in 1967 in reaction to behavioralist psychology.

Koestler's almost visceral defense of distinctly human qualities of emotion and judgment foreshadow some of what we may encounter when trying to understand the dual nature of cyberspace—both the mechanical and human aspects of it. Because it is a human creation it will embody all of the eccentricities, judgment, reason, sense, and dreams we consist of ourselves along with our flaws, weaknesses, and prejudices.

Almost 25 years ago, shortly after Koestler's novel was published, Robert Kennedy went to Detroit, a city that had recently been torn apart by a riot, and he spoke to how we measure the wealth of a community and the difficulty in quantifying intangible assets or values. He said,

> We cannot measure national spirit by the Dow Jones Average nor national achievement by the gross national product.
>
> For the gross national product includes air pollution and advertising for cigarettes, and ambulances to clear our highways of carnage. It counts special locks for our doors and jails for the people who break them . . . . The gross national product swells with equipment for the police to put down riots in our cities. And though it is not diminished by the damage these riots do, still it goes up as slums are rebuilt on their ashes.
>
> And if the gross national product includes all this, there is much that it does not comprehend. It does not allow for the health of our families, the quality of their education, or the joy of their play.
>
> . . . The gross national product measures neither our wit nor our courage, neither our wisdom nor our learning, neither our compassion nor our devotion to country. It measures everything, in short, except that which makes life worthwhile; and it can tell us everything about America—except whether we are proud to be Americans.

Similarly, we can look at the speeds of our telecommunications networks, the millions of miles of cable, fiber, and copper, ascertain the processing power of advanced computers, and measure their memory capacities. We can look, too, at the profit margins of software providers, movie studios, and record companies. In the final analysis, we can tell people everything about the state and quality of our network except those things that make use of such a network worthwhile. Whether the "console community" that is developing in cyberspace is one of enjoyment and wonder or the potential domain of digital derelicts who may pillage our community with acts of "electronic wilding" is the question Robert Kennedy would ask today.

We all have to remain cognizant of the fact that for all the glitter and gold out there on the frontier, there is, existing simultaneously, a

sinister side to cyberspace. It is an aspect of life that every community, whether real or virtual, has to deal with. So even as we look to
the network and to the first colonies on the electronic frontier to
empower human beings with the tools of the Information Age, to
improve people's lives, and to provide entertainment and enjoyment,
the potential for harm in the networked community may become
more than a "virtual reality"; it may become a real reality.

In 1989, I requested a comprehensive report from the General
Accounting Office (GAO) on how federal agencies use, obtain, verify,
and protect personal information; how individuals are made aware
of information collected about them; and what telecommunications
and network facilities agencies' systems use to transmit data. The
GAO reported that personal information at the federal level is maintained in about 2,000 program management, payroll, personnel, financial, and other types of systems. Data in these systems included
social security numbers, names and addresses, and financial, health,
education, demographic, and occupational information. The GAO
reported that although data in about 91 percent of the systems were
covered by the Privacy Act, many agencies still share the personal
information they maintain with other federal, state, and local agencies, as well as with the private sector.

The GAO study also indicated that the government obtains information electronically from third-party sources. Twenty percent of
the agencies surveyed reported that they collected personal information electronically from third-party sources, such as credit bureaus,
state divisions of motor vehicles, and insurance companies. This
study raises two issues for us. One, How does the government continue to protect the integrity of information collected about us for
legitimate purposes? and two, Is the government collecting only the
needed information?

The same GAO report noted that some of the government's largest databases of personal information are accessed remotely and electronically. Forty-five percent of the databases surveyed were accessed through the public switched network, such as through AT&T
or MCI, or through easily accessed commercial networks. The security of such systems is an obvious concern we will need to address on
an ongoing basis.

A more recent GAO survey, one I requested just last year, had to
deal with a proposal some of you may be aware of from the FBI to
meet its future wiretapping needs. In requesting the GAO investigation, I wanted to ascertain what technological alternatives to making
the entire telephone network—including computer equipment—"wiretap
ready" for the FBI were available. I also wanted to find out what

exactly the FBI's wiretapping needs were and what the estimated cost would be. The study concluded that in its April of 1992 legislative proposal, the FBI did not define its wiretapping needs adequately. No specifics and no cost estimates were available.

I can't tell you what alternatives to making computers, PBX equipment, and the telephone network "wiretap ready" may be available to the FBI because the Bureau classified that portion of the GAO report as national security information. I will tell you, however, that it is my personal belief that searching for alternatives to their current proposal is where the FBI should be focusing their efforts.

This is not only an issue of privacy; for U.S. manufacturers of telecommunications equipment, the FBI proposal and the NSA [National Security Agency] standard on encryption may also pose threats to the viability of their products on the international market. Because of my role as chairman of the Telecommunications Subcommittee, I felt obligated to investigate the issue thoroughly before any legislative proposals having far-reaching, or perhaps unintended, consequences moved through Congress.

I understand that the FBI considers wiretaps an essential information-gathering tool when fighting crime. But I am hopeful that some accommodation can be found because I feel strongly that we need to make our networks, databases, and terminal equipment more secure, not less secure, to invisible trespassers and others. The same is true for wireless encryption. The National Security Agency wants the industry to accept a standard that many believe is too easy to decode. As the telecommunications revolution goes wireless and telephone conversations, computer data, and business information increasingly travel through the air, we need to make sure privacy and confidentiality are protected to the maximum extent possible.

Let me get away from the government's side of cyberspace to some of the threats to personal privacy as they arrive from the private sector or from private individuals.

First, private industry. Erik Larson published a book last year called *The Naked Consumer*. In it, he unveils some of the tools Madison Avenue is employing to find out more about you and, subsequently, how they direct their sales pitch to you accordingly. New marketing technologies are being refined using sophisticated software that takes huge amounts of information collected from various sources and combines it into a single database. Larson calls this cross-referenced information "recombinant information." It can include court records, credit card balances, bank account information, magazine subscriptions, store purchases, and a "host of personal data collected discretely by companies of all kinds and then widely, avidly, and aggressively marketed to anyone willing to pay for it."

Larson notes in his book, for instance, that in 1988, one long-distance company began trading the names of its millions of customers. Anyone who rented the list could select customers who were female, who made international calls, or who traveled a lot. The author relates that the long-distance company tracked customers' travels through the use of the company's telephone calling card, designed to be used on the road.

Much of the collecting of personal information about Americans may be occurring anytime someone calls an 800 or 900 telephone number. Unbeknownst to most Americans, companies can receive the name, billing address, and telephone number of every caller to their 800 or 900 numbers. This information can then be reused, bought, and sold without restriction.

In the last session of Congress, I offered legislation to help combat this kind of personal information hijacking. I will introduce it again this session and will push for its adoption aggressively. It requires that recipients of personal information gleaned from the network during an 800 or 900 call NOT reuse or sell that information without receiving the *affirmative* consent of the caller first.

I'd like to read you a short excerpt from another book, this one by Jeffrey Rothfeder, called *Privacy for Sale*. By reading it, I think I'll give you all some sense—if you don't have it already—of how much information is readily available.

The author writes:

> I chose Dan Rather as my test case because I was told the stoic, tight-lipped CBS anchorman has taken numerous steps to guard his personal information. With this in mind, he seemed like the perfect subject to assess the limits of the [information] underground . . . . But in the end, it only involved extra keystrokes on my computer and patience. I started with Dan Rather's credit report because that's the simplest bit of confidential information to obtain.
>
> But to get that I needed his complete address and preferably—but not necessarily—his social security number. So I began by requesting from the superbureau Rather's credit report header, which contains biographical data taken from credit bureau files and usually available by just typing in a person's name . . . . Once I had Dan Rather's address and social security number, getting his credit report was easy . . . . It's hard not to be envious of a clean credit report like Rather's. . . .
>
> However, delving further into Dan Rather's electronic alter ego offers a slightly different perspective on him . . . . I obtained a list of the stores he shops at and how much he spends . . . . I learned

that Rather doesn't spend a lot on entertainment. Shopping seems to be more his speed . . . . He shopped at five clothing stores and ate at only two restaurants, both ethnic. Rather spent 10 times more on apparel than on dining out."

That's all I'll read to you, but you should know that it is just the tip of the iceberg of things the data superbureaus have on all of us.

As you can see, the Information Age is about more than just information. What we're communicating is not just raw data. It's values; it's meaning. It's a message about the very purpose of our lives. Communications has the power to change the way people live, the power to overthrow governments.

As networked communities in cyberspace develop further, the rights and responsibilities of all of its inhabitants, and all of us as well, need to reflect the underlying power they possess, in addition to the promise the network holds for us.

By looking at an individual's relationship with the machines and the network, we can also try to formulate boundaries of law in the network's borderless world that will serve well the community which inhabits it. In a sense, we begin with a sort of "technological anthropology" and work from there to the manifestation of human potentialities and needs in the technological milieu of the network as a whole. John Sculley of Apple Computer has spoken of developing "human-centric" technology to reflect his company's emphasis on development of information appliances for everyday people.

In 1991, Professor Laurence Tribe of Harvard Law School gave an address at the first conference on computers, freedom, and privacy, out in California, entitled "The Constitution in Cyberspace." The question he raised was how constitutional protections written two centuries ago remain intact in the context of a world, a digital world, the Constitution's framers could never have envisioned.

It is a fundamental question for us. As a policymaker at the federal level, I will be holding hearings on these issues, especially the electronic consequences to personal privacy in the Information Age. Big questions loom for us to answer—questions that will need not only legal, copyright, First Amendment, or communications law answers, but also societal and political answers as well. We will be breaking new ground in this increasingly interconnected world.

Here are some of the questions we will have to explore and attempt to find answers to:

• Are the fundamental values of our society so universal and enduring that they will not be threatened by the advent of new technologies or any new subcultures such technologies produce?

- Or will cyberspace become some lawless place, where the Constitution is cracked open by fiber fissures created when trying to convert a 200-year-old parchment into a binary world of zeros and ones? Can it continue to be a "living, breathing document?"
- Or will cyberspace develop its own distinct laws? Will it develop "digital vigilantes" to patrol and police the electronic bulletin boards and electronic highways? What indigenous political institutions may develop in such a vacuum?
- Could a closed system develop in the network with its own closed value system?

I agree with Professor Tribe when he stated that science and technology open options, create possibilities, suggest incompatibilities, and generate threats. *Yet they do not alter what is "right" and what is "wrong."* Because in the absence of relevant laws, egregious transgressions of what most of us perceive to be right and wrong can occur. This includes both unauthorized electronic trespassing and infringements upon free speech rights, as well as some of the tactics employed by hackers, "phone phreaks," and others.

Once we clearly define what the rules are, then we can deal with some of those digital desperados out there like the Dark Avenger—the Bulgarian notorious for spreading his computer viruses around the world. The most recent issue of *Discover* magazine, for instance, contains excerpts from a book which relates how the Dark Avenger has developed a virus that will mutate in 4 billion different ways, making it very difficult to vaccinate a system against. When the network goes global, how do we protect ourselves against cross-boundary access to personal data or infringements on our community? Will the privacy rules that govern U.S. systems be adhered to if the network or data is accessed from abroad? We need to build in such protections sooner rather than later.

I have also seen the magazines of the so-called "cyberpunk" community—magazines like *2600: The Hacker Quarterly.* A recent issue, for example, contained articles on how to steal long-distance service from a pay phone, how to defeat call-back verification, a program for a "simple C virus," and other articles indicating, quite explicitly, how to operate on the margins of, or to cross, what most of us would consider responsible behavior. Without appropriate laws that speak to our new, networked communities, the risk is that some of the citizens may be lost, not knowing exactly what their rights and responsibilities are.

Often, when engaged in the legislative arena, I find myself exasperated by representatives of certain groups or certain industries who appear to walk through life with blinders on—failing to see the larger forces at work or the proverbial "big picture." Similarly, there are numerous "geniuses" in the telecommunications or computer arena— self-described futurists and others—who appear to me to walk through life with binoculars on. They can see way out into the distance, yet anything in the immediate future is completely out of focus.

Luckily there are some who do see the immediate, pragmatic steps we need to take, for the good of the network from a technical sense, but who also remain cognizant of the possible pitfalls nearby from a societal perspective—people like Mitch Kapor and those at the Electronic Frontier Foundation, for instance, Gene Kimmelman and Mark Cooper at the Consumer Federation of America, and Jan Goldman from the ACLU's Privacy and Technology Project. And there are others in this room as well. I look forward to working closely with all of you in the future on both the technical, regulatory side of cyberspace and on the societal, privacy implications as well.

As I said earlier, it is important to do this now. To not look at these issues is running the risk of logging on one morning and entering the vast emptiness of a monadic realm that perverts our hopes for a true community with invasions of privacy and digitized demonstrations of injury and violence.

To make our vision of a networked community a reality, we have to remember the core values behind the Communications Act of 1934— values which apply as well in cyberspace as they did in the New Deal. One value is to ensure universal access to every person in our country—rich or poor. Another is to ensure diversity:

- That there is a multitude of "media tongues" that can speak;
- That you don't have to work for the biggest and most powerful companies, or be a certain kind of person, to get access;
- That the *smallest voices*, those articulating *creative* ideas, those with information to communicate, have access to the telecommunications network;
- That they can be free and strong and separate from the larger voice that may want to be more monopolistic and drown them out, and that localism can be, in fact, fostered through this network; and
- That it's not just a couple of voices coming from New York or Los Angeles, but voices all across the country that can use the telecommunications system.

For this reason I am advocating that America needs to take an interim step on the way to a fully functional broadband system for the country, a step that will avail a great number of people the opportunity to access a little bit of the future today, through digital telephone service to the home.

The power of communications is the power to help us learn more about the world and to bridge the gaps that separate our differences. I would like to work with all of you to make that kind of network a "real" reality.

Thank you for inviting me to speak with you tonight.

Biographical Sketches

STEERING COMMITTEE

Dorothy E. Denning (steering committee chair) is a professor of computer science at Georgetown University, where she is currently working on policy and technical issues relating to cryptography and wiretapping, and has served as an independent reviewer of the government's escrowed encryption system. Before coming to Georgetown in 1991, she was a member of the research staff at Digital Equipment Corporation's Systems Research Center, a senior staff scientist at SRI, and an associate professor of computer science at Purdue University. She is author of *Cryptography and Data Security* and numerous papers on information security, and in 1990 received the Distinguished Lecturer in Computer Security Award. Dr. Denning is chair of the International Cryptography Institute sponsored by the National Intellectual Property Law Institute and cochair of the Association of Computing Machinery's Conference on Computer and Communications Security. She is past president of the International Association for Cryptologic Research. She received a Ph.D. degree in computer science from Purdue.

Anne Wells Branscomb is a member of the Center for Information Policy Research at Harvard University. She is a communications lawyer with practical experience representing broadcasters, cable television companies, database providers, semiconductor chip manufacturers, and publishers of newsletters, newspapers, and books. Ms. Branscomb is an honors graduate of the George Washington University Law School, holds degrees in political science from Harvard University and the University of North Carolina, spent a year as a visiting scholar at the Yale University Law School, and studied international relations at the London School of Economics as a Rotary Foundation Fellow. She has served as chair of the Communications Law Division of the American Bar Association's Science and Technology Section. She is also a member of the U.S. Department of Commerce Technical Advisory Board, a trustee of EDUCOM, a member of the Commission on Freedom and Equality of Access to Information, a trustee of the Pacific Telecommunications Council, and a contributing editor to the *Information Society* and the *Journal of Communication.*

Mitchell D. Kapor is cofounder and chair of the Electronic Frontier Foundation, an organization that works to develop and implement public policies to promote openness, diversity, and innovation in emerging electronic social environments. He is also chair of ON Technology, a developer of local area network applications for collaborative computing. He received a B.A. (psychology, linguistics, and computer science, 1971) from Yale College and an M.A. (psychology, 1978) from Beacon College, and studied management as a postgraduate at the Massachusetts Institute of Technology's Sloan School of Management (1979). Mr. Kapor is founder of the Lotus Development Corporation and served as its chief executive officer and president (1982-1984), and also chair (1984-1986). He is the designer of Lotus 1-2-3, Agenda, and many other software applications. Mr. Kapor is the chair of the Commercial Internet Exchange (CIX), a not-for-profit association involved in the development of arrangements and facilities that connect independent networking carriers into a global information infrastructure. Mr. Kapor served on the Computer Science and Telecommunications Board of the National Research Council. He is also an adjunct research fellow at Harvard University's John F. Kennedy School of Government in the area of information technology policy.

Stephen T. Kent is chief scientist, Security Technology, for Bolt, Beranek, and Newman Inc., where he works with commercial and government clients to develop solutions for network and computer security problems. He served as a member of the Internet Architecture Board from 1985 to 1994, and chairs the Privacy and Security Research Group of the Internet Research Task Force. Dr. Kent served on the Presidential SKIPJACK Review panel and various National Research Council technical panels, and was a member of the board of directors of the International Association for Cryptologic Research from 1982 to 1989. He received the S.M., E.E., and Ph.D. degrees in computer science from the Massachusetts Institute of Technology. He is the author of numerous articles on network security and has lectured on the topic throughout the United States, Europe, and Australia.

George M. Perry is vice president and general counsel for Prodigy Services Company, which he joined in 1984 (then called TRINTEX); he has oversight of issues relating to the rights and responsibilities of a commercial electronic information and transaction provider. He is also responsible for tracking federal and state regulatory policies as they relate to PRODIGY's provision of competitive services and technologies. He received a B.A. (1961) from Columbia University and a J.D. (1964) from the University of California at Berkeley. Mr. Perry is currently concentrating on issues and policies that impinge on the delivery of information services to the general public in the constantly changing technology and regulatory environment.